The Soul of Os

The Place of Mind in Early Osteopathic Life Science

Including reprints of Coues' <u>Biogen</u> and Hoffman's <u>Esoteric Osteopathy</u> with contemporary explanatory prefaces by

Zachary Comeaux DO

BookLocker.com, Inc.
2009

For additional copies, go to Booklocker.com or zacharycomeaux.com

This work is dedicated to Mom, who, as I write, is teasing us with the question of what is the limit of life.

Other books by the author:

Robert Fulford DO and the Philosopher Physician

Fire on the Prairie: the life and times of Andrew Taylor Still, founder of Osteopathic Medicine

The Healer's Heart – a chronicle of the inner journey in meeting Life's challenges

Harmonic Healing – a guide to Facilitated Oscillatory Release and other rhythmic myofascial techniques

Contents:

Introduction

Though not part of the business of everyday practice, osteopathy from the beginning, as part of its scope and focus, has included the definition and understanding of the human person. This included the question of "What is Life." This philosophical theme is reflected in the words of A. T. Still,

"No one knows who the philosopher was that first asked the question, What is life. But all intelligent people are interested in the solution of this problem."
1

The question arose in the context of Still's appreciation of social and scientific advances of his century and his desire to apply science in the osteopathic clinical context of improving quality of life and health.

Throughout human experience there are two intertwined efforts in the quest to reconcile science and philosophy. The one asks us to accept our material character as the complete description of who we are by nature, and one views man as additionally possessing a soul which draws him to a higher purpose, to strive to be more, to fulfill a destiny. Classically we see this latter theme set down in Greek mythology in The Odyssey, and the myth of Prometheus. The Genesis account of the fall from the Garden of Eden reflects the intricacies of this same theme. In early Christian times the competition for a proper place for Man in the universe was reflected in the orthodox approach to faith versus the Gnostic belief that salvation came from a certain type of esoteric, or hidden, knowledge.

In the field of science, Einstein's wrestling with the relationship of light, energy and matter attempted to expand our understanding of the physical universe, a modern quest for the Holy Grail. In our own day the Human Genome Project and the quest for a cancer cure may reflect this same dynamic to understand the big picture.

In all this searching there arises a dilemma between the acceptance of tangible reality as the completeness of existence, or on the other hand, the allowance for the reality of unseen, or unrecognized, forces. The struggle has been defined in the past as the dialectic between the vitalist and materialist point of view. Throughout the historical progress of the field of inquiry which acknowledges the legitimacy of the unknown, there have been the Essenes, the Gnostics, the followers of Paracelsus, Rosicrucians, Masons, Theosophists all claiming to possess uncommon knowledge of the unseen behind everyday reality.

Dr. Still, founder of osteopathy, placed himself in the company of these searchers. He associated as a freemason, although his independence of thought made him a renegade there, too, and he was sometimes challenged in his loyalties. In an editorial in the Journal of Osteopathy in 1901 he defends the practice of Masonic secrecy, but in the context of religious tolerance. He ends the editorial with:

Introduction

"It is much to be hoped that we will some day have something better than masonry, and that the church from Mahomet down will give way to something better, and all rally around the flag on whose face you read ` Love thy neighbor as thyself'."

Still uses the name of God hundreds of times and praises Christianity. Why did he not profess a Christian church affiliation, especially as the son of a Methodist preacher?

Still's unorthodox view of health and healing led him to be rejected by the denominational churches, in fact to be denounced from the pulpit.3 His early practice struggled under the weight of this stigma.

Since Still's proclaiming osteopathy in 1874, the esoteric theme and the contested boundary between physics and metaphysics has been an element of the quest to define the scope of osteopathic practice. Still several times describes osteopathy as the science of mind, matter and motion. Still would reflect on looking to discover the health in the patient by the correct relationships between tissues. This addresses the matter and motion components of his science, but what of the mind? If one reads carefully, one finds Still interpreting anatomical finding and relationships as reading the mind of God in nature, in the anatomical design of the patient.

This rather lengthy quote, also from the 1901 Journal of Osteopathy, captures these relationships better than I can find in his lengthier texts:

"What is God? If all of man, with his mind, matter and motion, is one being, what is the universe but a being? It has mind, matter and motion. It does its work well and wisely, still it is only one universe. Then mind to the universe is the same that mind is to man. This God would be the universe. We are in the universe therefore, we are with God and help compose that great all, and journey as it journeys. That great composite is eternal, and so are we. We have lived, do live and will live out the full number of days of the universe. Thus to us a universe means all space and all therein contained. This signifies the universal universe. Man under the same law of reasoning would be a dependent universe, while the universal universe is not dependent, because it is the all of all, specially and universally, mental, motor and material. The individuality of mind with its independence of all else, to me seems to be impossible, because of the superior endowments of the mind over motor, which motor is above material in quality, but not at all its superior mentally. Thus both the physical and mental submit to the higher principle, which makes a unit of the three; matter, motion, and mind. Thus the universe is a being, with the mental, motor and material combined, and leaving the management of all under the mental. Thus we have God as mind in union, working in union with the motor and physical." 4

However, the conventional way of interpreting the essence of osteopathy is as if Still were solely directing us to study our physical anatomy. And he does drive us to study anatomy. However, there is more to following Still in the development of his practice than anatomy in the conventional sense. Still derived some information from study, but he was also described as a telepath. Furthermore, he was reported to maintain contact with a medium, Matah, the Indian.

Most early osteopathic texts claimed to be following and represent the content of Still's teaching by describing osteopathy primarily as an organized system of procedures. Barber, McConnell, Hazzard, all describe osteopathy primarily from the point of view of technique. Indeed, little of that writing reflects the broader quest to explore the nature of life and man's participation in it.

Still himself explores that theme in the chapter in the Philosophy and Mechanical Principles of Osteopathy entitled "Biogen". He describes a life force in the context of a dualistic approach to the nature of man.

"All material bodies have life terrestrial and all space has life, ethereal or spiritual life. The two, when united, form man."

And later:

"A man, biogenic force, means both lives in united action...Thus we say biogen or dual life, that life means eternal reciprocity that permeates all nature." 5

But how did Still come to conceptualize the human experience in these terms? Still's style was rarely to reference his sources. For a frontiersman, the scope of his written discourse seemed scattered; however, to the astute investigator it demonstrates that Still was either well read or otherwise well informed as to the scientific and philosophical ideas of his days. His apparent ramblings are in actuality his best attempt at synthesizing a broad spectrum of complex ideas presented by a host of thinkers.

The current book is to be read as a companion to Still's *Philosophy and Mechanical Principles of Osteopathy*. It includes a reprinting of the full text of Coues' *Biogen; a Speculation in the Origin and Nature of Life* (1874) which predates *Philosophy and Mechanical Principles* (1992) and in someway was no doubt Still's source of the term.

Who was Elliot Coues? What was his contribution to nineteenth century science? How did Still access his work? The introduction to this text will deal with this and related issues.

This book includes a second frequently overlooked work, Herbert Hoffman's *Esoteric Osteopathy*. Hoffman writes directly on the topic of Mind and its role in osteopathic treatment. I had mentioned Still's having contact with a psychic medium or guide. Still wrote of this in his notebooks which have subsequently been destroyed by his family to avoid an appearance of mental unbalance or eccentricity. 6 Interestingly enough, Hoffman's small book, from the early era of osteopathy, also showing Indian influence.

Who was Hoffman's mentor in regard to this subject? Was it Andrew Still?

Osteopathy has over the years evolved a panoply of techniques and philosophical points of view, from

Introduction

stressing biomechanical approaches to more subtle
work. Often these are seen as contradictory or at least
competing as the true paradigm. The current work
seeks to bring forth some little known evidence that
the osteopathic approach to health was based on an
understanding of the need to bridge these two
paradigms, not to establish a competition between
them.

How relevant are these topics? Today we wrestle with
the frontier between pharmacologic and alternative
medicine, between bioscience as self-organization of
matter, with consciousness as an emergent property of
biomolecular complexity, versus a creationist belief in
intelligent design. Twenty-five hundred years ago the
philosopher commented that there is nothing new
under the sun. How true is this today?

As Part 3 I have included an essay of my own, written
about ten years ago but as yet unpublished. It revisits
the long view of the vitalist/materialist controversy but
brings it into era of proteomics and genetic
determinism.

In a more practical vein, we know that mentation,
everyday social and mental activity, is not uniform
throughout life. Certainly our early developmental
phases reflect this thought. But more and more, as we
become more sophisticated with life support and
resuscitation, the question of "what is life" has a more
poignant application for many of us. Cerebrovascular
accidents, temporary hypoxia, or progressive dementia
leave millions alive yet without "normal" mental life
and function. How to make sense out of this phase of

existence is a quandary. The question of "what is life" bears on the issues of abortion/right of choice, euthanasia, as well as conventional end of life care. If you have had or do have clinical or personal experience with these situations, as you read see if the thoughts of these authors passed on here contribute any clarity.

Between the time I have written the dedication and now finished the introduction I have spend the last two days of my mother's life communicating with her in an altered state. As with my father's passing, nothing can convince me that conventional cognition and consciousness are the sole determiners of meaningful human life. Once out of touch, we presume an individual to be in a sub-human state. My experience confirms for me that death does not represent deterioration or termination of personal life as much as a commencement of some other phase of life.
Still suggested such when he referred to the human body as the second placenta.7

1. Still, A Philosophy and Mechanical Principles of Osteopathy, p 249
2. Journal of Osteopathy in October, 1901, (p. 317)
3. Still, C Frontier Doctor, p. 41
4. Journal of Osteopathy in October, 1901, (p. 317)
5. Still, A Philosophy and Mechanical Principles of Osteopathy, p. 251
6. personal communication, Jerry Dickey DO
7. Still, A Osteopathy: Research and Practice, Eastland Press, Seattle, 1992, p. 11 from original pub. 1910 by Author, Journal Printing Co., Kirksville, MO

Part 1

Biogen: a speculation on the origin and nature of life

Contemporary Preface to Biogen

There are few guides for the student who looks beyond the anatomic/biomechanical interpretation of osteopathy according to Dr. Still. However, one is immediately excited, if one has been intrigued by Still's description of Biogen in his wrestling with the concepts of body-mind- spirit, when one discovers Coues. Elliot Coues' publication (1884) certainly predates Still's publication of *The Philosophy and Mechanical Principles of Osteopathy* (1901). Coues lays claim to credit in coining the term biogen. Since Still would not cite references, we know little directly about how Still came to use the term, biogen, in approaching the problem of "what is life?"

Did they have direct contact? Kirksville was on the active rail line and a busy center of therapeutic activity. Once he left Virginia as a youth, Still's travels eastward were limited to his participation in the 1903 Columbia Exposition in St. Louis. By the time of his writing of his monograph on Biogen, Coues, as we will see in reviewing his biography, was ensconced solidly in the Washington intellectual community. His days as an explorer of the American West as an army surgeon and scientific field officer were behind him.

An analysis of the text does not help greatly. On an initial reading, there seem to be minimal conceptual links. The setting and occasion for the use of the word is different. Still's eclectic style in writing, or compiling, his books places chapters in random juxtaposition creating a literary collage. Still's Biogen chapter rests between Fevers and Smallpox. Still begins his chapter on Biogen by setting the stage to help osteopathy fit in with nineteenth century modernity and rapidly expanding technological and scientific progress. He ends it with reflections on the negative effects on man's evolutionary development of war as the slaughter of the fittest.

Coues, on the other hand, is working in dialogic fashion with several previous presenters to the Washington Philosophical Society in reviewing and refining the perennial vitalist/materialist argument on the divide between physics and metaphysics. The main body of the text is written in Socratic fashion with an appeal to common experience and current scientific ideas, including the now quaint but then advanced concept of luminiferous ether, the medium for transmission of light. He does so with an urgency to call for balance in the face of the rapidly rising shift in science to limit thought related to biology only to physico-chemical considerations. He sees the scope of biological science to be much broader. His analysis is of necessity more rational and organized than Still's. In his appendix, he is more direct and succinct in describing his view of the relationship between body, soul and spirit.

One could assume that with this rather loose congruence between the two writers, beyond the common use of the word, that Still picked up the concepts and topic from discussion with a third party, such as his old friend and intellectual mentor, Colonel James Abbott. Abbott was born and educated in the East and shared Still's interest in science and philosophy. They encountered one another as members of the Kansas Legislature, militia, and in founding Baldwin College. There would have been plenty of occasions for intellectual discussion and Abbott is cited by Still in his *Philosophy of Osteopathy* among a short list of major influences. 1

Beyond this scope of inference, there are some details and key phrases in the texts, as we will see below, which may suggest that Still had encountered a copy of the written presentation of Coues' *Biogen: a speculation on the origin and nature of life*.

Coues career

Although he began and ended his intellectual career in the nation's capital, Coues was not a salon philosopher. His early love of nature led him at seventeen to sign aboard the schooner Charmer to study maritime wildlife in Labrador. Encouraged by his lifelong patron and mentor, Spencer Baird, assistant secretary of the Smithsonian Institution, Coues would become one of the foremost of American ornithologists. Birding, in those days was done by the search, shoot, inspect, skin and classify method. Coues' *Key to North American Birds* includes taxonomy but also instruction on preparing specimens, and principles of biology, included his theory of biogen.

The real drama of Coues' life unfolded as he served as a medical officer and field biologist for the U.S. Army in Colorado and Arizona. Identification of new species was done with an eye out for Apache Indians anxious to retain and regain their home territory. Serving in a role similar to Lewis and Clark, Coues medical duties were minimal and he regularly shipped hundreds of new species east, including new species or sub-species of birds and mammals. His related publications fill volumes.

Of more interest to us here is his affiliation with Freemasonry which began in 1867 during a deployment in South Carolina. This would reflect a sensitive but philosophically variant view from orthodox religion, of which he was known to be critical. However, it is not until his address to the Washington Philosophical Society, the transcript of which we have here in this book, that Coues makes a definitive break with conventional science of his day. After this date, he became more expressive of his broader world view. He became acquainted with Madame Helena Blavatsky, the popularizer and charismatic founder of the Theosophical Society, an eastern/western syncretism of occult practice. Coues was, in 1885, to become the president of the society, also serving in the same period as the censor, or literature reviewer, for the American Society of Psychical Research. During this period he was a prolific writer and frequent conference speaker on topics related to spiritual science and related topics. This, and the Freemason connection he shared with Andrew Still, could easily have made his thought visible in print to an exploring and curious Andrew Still.

Few in the spiritualist circles appreciated the effort Coues made to reconcile the apparently esoteric sphere with that of empiric science. So, later in life, he returned to the field of biological science and the history of western exploration, writing biological definitions for the Century Dictionary and material for the Encyclopedia Britannica before becoming involved in the controversial reediting of the Journals of Lewis and Clark, of Zebulon Pike as well as early French and Spanish explorers.

Gist of Coues' Biogen argument

Elliot Coues takes issue with the emergent perspective in bioscience that looked at protoplasm as the basic element and condition of life. It had been proposed by advancing science that life is concomitant with protoplasm, formed spontaneously and advanced by the progressively complex aggregation of chemical elements, for us the "emergent property" hypothesis.

"Biology has proven that Life was a mode of matter and motion, ergo, protoplasm was the life principle; and it has just upon the point of being discovered by the Society, when the protoplasm, which the Society had examined, died. So the vital principle had given them the slip."

Coues makes a different proposal. Existential awareness of oneself as conscious and operative is the motivation for Coues to speak. He develops an argument to align one's consciousness as an operative power and reference point which in itself, because of its cogency, is self-validating in a way that the regional succession of chemical actions can never be. In so doing, he reaffirms that this critical self-consciousness is both evidence of an undefined vital force, which can be potentially evaluated by science, in much the same way as science of the times was coming to accept the "luminiferous ether" as a way of understanding the phenomenon of light. At the same time, he cites that this universal cogency of the self-awareness of human kind is linked through the will to motion- all motion - and to Universal Mind. Simultaneously, he uses this self evident conviction to reaffirm the perennial and universal phenomenon of faith, including faith in the ultimate and supreme consciousness, God, as an essential element of scientific cosmology.

Although this is Coues' main argument, Still would be compatible with it but does not cite it in the development of his own thought. However as mentioned before, there are key phases on a more physical, less psychological basis, which bind the two presentations together and hint at some contact by Still, at least with Coues' written text.

From Still
"We see the form of each world, and call the united action biogenic life. All material bodies have life terrestrial and all space life ethereal or spiritual life. The two, when united, form man." 2

From Coues
"….spirit…I have postulated that it actually does exist, and defined it as self conscious force (biogen). I have speculated that a living body results from the action of spirit on matter and that life subsists on the union of the two." 3

From Still

"Thus we say biogen or dual life, that life means eternal reciprocity that permeates all nature." 4

Motion

Both authors refer to the aspect of life represented by motion. Each reflects a familiarity with the philosophy, physics and cosmology of Herbert Spencer, an important philosopher of that era. Coues make the argument however, that there needs to be a *cause* for motion, or a moving system, calling on the Aristotelian concept of a First Cause. And so:

"Force cannot act where it is not; neither can it act with nothing to act upon; its presence and operation upon matter are, therefore, necessary conditions of its manifestation; all the manifestations of life are ultimately resolvable into modes of motion, and in the particular modes of motion exhibited by living things, *and by no others,* are evidenced the presence and operation of the vital principle, the energy of which differs from other energies precisely as the modes of motion of living things differ from those of all things that do not live."

A familiarity with Spencer is also disclosed by the reverence which the two men have for the *unknowable*.

Spirit

Still's sense of spiritualism and telepathy, though attested by family members and friends, are hidden in his language. Echoing his previous work Still cites:

"First there is the material body; second, the spiritual being; third, a being of mind which is far superior to all vital motions and material forms, whose duty it is to wisely manage this great engine of life." 6

He gives little detail regarding the sense of the spirit per se, recapitulating the argument for a "celestial force" underlying life processes. 7

But Coues is concentrating on a different argument, the relevance of which should be apparent in our day. He was concentrating on a rebuttal to the position that protoplasm, the simplest evidence of life, was proposed to be what we call today an emergent property of self-organized matter. The chemico-physical explanation for this proposed that physiologic motion, concomitant with life, was a result of the instability occurring with larger and more complex aggregations of matter. The Darwinist point of view of survival of the most adaptive carried the rest of the argument. Coues most fundamental reply to this is to cite the impossibility of the random occurrence of a system which could then self-replicate. To say that a biological system could spontaneously organize in a way that continued by laying its own egg was a scientifically untenable, and non demonstrable, position.

Especially in the appendix, as you will see, Coues develops a psychology/cosmology which related his concept of biogen, spirit, matter, soul in such a way as to suggest that biogen is an extended capacity of matter, and not really a dualistic way of viewing the world, the usual vulnerable attribute of a vitalistic theory.

Both Still and Coues invoke the importance of Mind in living organism, Coues by recognizing that awareness itself is evidence of a primary vital principle reflecting the Universal Mind. Still in turn refers to the study of anatomy:

"Let us accept and act on it as true, that life is that force sent forth by the Mind of the universe to move all nature, and apply all our energies to keep that living force at peace, by retaining the house of life in good form from foundation to dome. Let us read a few lines in the book of Nature. If we stop blood in transit…"

Coues states, in his appendix, that "The mind is the result of interaction between spirit and matter." And again:

"Whence emanated matter in the beginning is inscrutable; from nowhere, certainly, -if not from the self-conscious, self-determining universal Mind which willed to become so manifested." 9

In the case of both authors, linkage is made through this reference to mind to the third work discussed,

Esoteric Osteopathy, by Herbert Hoffman discussed as part two of this current book.

Current thought

So what is the relevance of this discussion for us today? In the not too distant past, Robert Fulford, DO, taught and operated under the presupposition that "thoughts are things" and had an impact for good or ill on the etheric or energetic body. His treatment relied upon the use of intention or mental activity as an operational principle capable of potent interaction with the patient. He had read Reichenbach as had Coues some generations before. 10

In the time since physics has advanced in its understanding of light and biochemical bonds, these issues seem archaic. However, the field of condensed matter physics is today a viable discipline as is the field of prebiology involved with the evolution of macromolecules and genetically orchestrated self-organization of molecular systems, let alone galaxies. The term intelligent matter seems to counter Coues argument for a distinction between mind and protoplasm. Coues would side with what we now call a creationist world view. Contemporary attempts to reconcile the vitalist/ materialist conflict include the concept of theistic evolution and intelligent design.

The contemporary study of the physiology of consciousness has moved beyond the discovery of nuclei, brain waves and neurotransmitters to the study of oscillatory synchrony and phase changes both among and between populations of neurons. Not only in the brain is this recognized, but oscillatory function is held by a large number of physiologists to be the means of information transmission and coordination in living systems in general.

Yet the questions are still debated. Is consciousness an emergent property of complex adaptive systems of genetically elaborated proteins alone? Or is there an extrinsic or participatory aspect of psychic life which is generated by something beyond the material as we know it?

Search the web and you will find that the questions and hypotheses of Still, of Coues, and of Hoffman are still hot topics today.

References:

1. Still, A Philosophy of Osteopathy, Indianapolis, IN, American Academy of Osteopathy, 1977, orig. by author Kirksville MO 1899 p.14
2. Still, A Philosophy and Mechanical Principles of Osteopathy, Osteopathic Enterprises, Kirksville, MO 1986 orig. Hudson-Kimberly Pub. Co. Kansas City, MO 1902 p. 251
3. Coues, E Biogen, a speculation on the origin and nature of life, Estes and Lauriat, Boston, 1884, p. 45
4. Still op. cit. 251
5. Coues, E op. cit p. 38
6. Still, op. cit. p.17
7. Still, op. cit. p. 252
8. ibid. p. 101
9. Coues, op. cit. p. 57

10. Comeaux, Z Robert Fulford DO and the Philosopher Physician, Eastland Press, Seattle, 2002

Other Resources:
Brodhead, M A Soldier-scientist in the American Southwest, The Arizona Historical Society, Tucson 1973

Cutright, P, Brodhead, M Elliot Coues, naturalist and frontier historian, University of Illinois Press, Urbana and Chicago, 1981

Biogen:
a speculation on the origin and nature of life

by Professor Elliot Coues
member of the National Academy of Sciences; of the American Philosophical Society; of the Philosophical and Biological Societies of Washington; etc., etc.

"As thou art fitted to receive it, so shall the light be given to thee."
The Demon of Darwin

Boston
Estes and Lauriat
1884

Biogen

Second edition

University Press
John Wilson, and Sons
Cambridge

To

William B. Taylor

Lately President of the Philosophical Society of
Washington, Learned in Science, Wise in
Philosophy, Faithful in all Life's Relations

This volume is inscribed
with respect and friendship

by

The Author

"The most general truth, not admitting of the inclusion in any other, does not admit of interpretation. Of necessity, therefore, explanation must bring us down to the inexplicable. The deepest truth we can get at must be unaccountable."

H. Spencer

"He who supposes, therefore, 'that the information of the senses is adequate (with the aid of mathematical reasoning) to explain phenomena of *all kinds,'* and refuses to admit 'that there are physical operations which are and ever will be incomprehensible to us,' betrays a very imperfect idea – no less of the impassible limitations of finite intellect, than of the fathomless profundity of Nature's system. He who thinks that by formally repudiating the mysterious, and confidently discarding the unknown, he thereby abolishes or in the slightest degree diminishes his insuperable nescience of the ultimate, - but imitates the ostrich, and deludes himself."

W.H. Taylor

Preface

In the spring of 1882 I was honored by an invitation, which I did not feel at liberty to disregard, from the President of the Philosophical Society of Washington, to address that learned body upon the general problem of Life – Whence, What, How, and Why.

The fascination of these questions, perpetually asked and unanswered, is due to the fact, that we know them to be unanswerable, yet feel that they will be answered somewhere, somehow, sometime, by every human being, each for himself.

The situation at the Philosophical Society I was given to understand to be this: The retiring President had in his last address discussed biology, contending that a certain "vital principle" caused Life, or was at any rate necessary for the purpose of Living. This would seem to be a reasonable proposition; but it has been regarded as more or less unphilosophical or unscientific, because the Society had not succeeded in finding out what the vital principle was, or indeed, where to find it at all. Mathematics had failed to find it at any point in the known dimensions of space. Physics had failed to find it in any kinesis of attraction and repulsion. Chemistry had failed to find it in any atomic or molecular combination. Then Biology – "The Science of Life"- had come to the rescue with a substance known as protoplasm; for Physics had proven that nothing existed but matter in motion; Biology had proven that Life was a mode of matter in motion; *ergo,* protoplasm was the vital principle; and it had been just upon this point of being discovered by the Society, when the protoplasm, which the Society had examined, died. So the vital principle had given them the slip, and the Physico-chemical Theory of Life had been unable to recover the same. It having thus become evident that there was a difference between something alive and the something dead, the "previous question" had obviously recurred.

I prepared what I had to say on the subject to the best of my ability, and carried it to the Society with much misgiving. For I could not say what I truly thought- and what else should any man say? – without introducing strangers to a select body of Washington scientists – such as God, Spirit, Soul, as factors into the problem of Life. Trusting, however, that their names are known, at least, I delivered the address subsequently entitled "Biogen".

No one who has frequented scientific societies can have failed to observe how naïve and natural are our exhibitions of human nature. We "elder children" cannot be outdone by the youngest in our harmless vanities. When some one is speaking, for example, we who are listening are busy with our pencils and note-books. To put down the best things he says? To put down the good things even? Why should we? These things take care of themselves, do they not? We watch him like a hawk, to pay ourselves for having to listen; to catch him tipping, and find fault with him afterwards, and have an excuse for speaking ourselves. We are all too full of our own ideas to listen to any one's else for any other purpose, or to any other terms. We immediately rise to complement the speaker with the most glittering generality, before confounding him with the utmost peculiarity. What could be more simple, more natural, more human, more childlike?

On the occasion to which I refer, for example, a philosopher said that he had listened to the, etc. addresses of the etc., with greatest, etc. But the speaker had adduced the consensus of mankind in support of his views, and the consensus of mankind was demonstrably erroneous in many particulars. For example, take the rainbow, which mankind had for years believed to be set in the sky by the Deity, as a thing of beauty, and a token, and a promise. Whereas the triumphant progress of modern science had shown its shape to be due to the circular equality of angle in the locus of the water-spherules, and its color to the varied refraction of light. For the rest, he could only refer the speaker to the well-known properties of protoplasm, and the modern theory of evolution.

A philosopher, waiving the usual opening formula, stated without reserve that there could not be anything in anything he had heard me say, because nothing existed but matter in motion.

A philosopher said that he could not imagine how the speaker could seriously ask such a questions as, What is the difference between a dead Amoeba and a live Amoeba? He should be almost ashamed to be called upon to answer such a simple question. From his manner I gathered that he wished I had asked him something hard.

A philosopher hoped that Professor Coues is not teach such heresies at the college where he habitually lectured.

A philosopher of an inquiring turn of mind, apparently, said that I had spoken of "soul" and "spirit" as of things whereof a man might possess more, or less; but that, if so, my views would remain without scientific basis until the invention of a "biometer" to measure the cubic contents or avoirdupois of a man's soul-stuff. Upon which I could not help thinking, and saying, that an instrument for measuring the soul should be the last thing some philosophers should wish to see invented – and applied.

A different kind of a philosopher spoke for a few moments; I will not transcribe his remarks. Our eyes met, and I new he understood me. But the pertinence of most of the remarks which followed the delivery of "Biogen" must be left to the reader to discover, upon perusal of the published minutes of the meeting (see Introduction). The general sense of the meeting was probably reflected in the remarks made privately to me by one of my friends: "Damn good English, Coues, and damn poor sense. You ought to get to be a good square flat-footed atheist, and then you won't take these fits."

When the question of publishing "Biogen" came up, I asked the advice of one who I knew would endeavor to dissuade me, in order to learn his reasons. He begged me not to publish it, for my own sake, because it would 'injure' your scientific reputation. Acting upon this advice, and wishing to discover, if possible, how an honest expression of honest convictions on any subject could injure any one's reputation for anything excepting insincerity, I immediately printed a small edition which was speedily exhausted.

The treatise having found favor in some eyes in whose penetration I have confidence is now republished without other changes than the addition of this Preface, the following Introduction, an Appendix, and some footnotes here and there. Should the line of thought presented be found to lead, or even to tend, in the right direction, it may be followed up hereafter; the Author being now in position to express himself more fully, freely and explicitly on the subject than he was when "Biogen" was first published.

Living as I have been for many years in a scientific atmosphere in which atheism and a very crass materialism are rife, as the fashionable foibles of many men otherwise really great, who almost hide their folly with their erudition, their good sense, their thousand manly and humane qualities, I am often told by scientists that they have no souls, and expect to die like dogs. What can I rejoin to such declarations from such sources? To such a one I can only answer evasively, that he must know his own nature, and probably destiny, better than he can expect me to; and that if he thinks he has no soul, and is to die like a dog, I have no means of proving him wrong; but that, speaking for myself alone, I know that I have a soul, and that I shall not die like a dog, because it is the nature of the soul God has given me to know its immortal self with a kind of knowledge in comparison with which the knowledge of material things acquired by the bodily senses is no knowledge, but delusion only- with a kind of knowledge whose servant, not whose master, is reason- with a kind of consciousness which is self-conscious.

If my philosophy approves this consciousness, if my science supports and strengthens it, I am happy. If they do not, of what use are they to me? Idle, wasteful slaves, that eats into the life and substance of their master – not worth their keep.

Not many men, I fear, think; it tires them, and hurts their feelings; it strains their constitutions; a more or less sequential series of bodily sensations is an easier way of life, that "embarrassing predicament which precedes death," and saves the trouble of thinking. A few men think, and their hard thinking hardens the brain, and sets it in a mould, and no thought of another shape can find fit or rest there. And the spider of vanity spins her web there, and nimbly transverses its geometric threads, and lo! A system of philosophy. But such shall pass also, brother philosopher; your science and mine must bend the knee to our common humanity, there to learn that knowledge is not wisdom till it becomes self-knowledge, nor this masterful till it has mastered self. *Then-* forge the chains of your systems as you may; the veriest gossamer thread shall be stronger to bear you up than they to hold you down.

Introduction

(Extracted from the Bulletin of the Philosophical Society of Washington, vol. v, pp. 102-105)

217th meeting   May 6, 1882

President Wm. B. Taylor in the Chair

The first communication was by Mr. Elliott Coues,

On the Possibilities of Protoplasm

The following is an abstract of this communication which has been published at greater length under the title – Biogen: a speculation on the Origin and Nature of Life. Abridged from a paper on the 'Possibilities of Protoplasm', read before the Philosophical Society of Washington, May 6th, 1882. Washington: Judd and Detweiler. 1882 8 vo, pp. 27

"Referring to the previous papers on the subject of Life, my Mr. Woodward and Mr. Ward, the speaker opposed any pure physico-chemical theory, and adhered to the doctrine of the actual existence of a 'vital principle'. Granting that all substances, including protoplasm, have been evolved from nebulous matter; that evolution to the protoplasmic state is necessary for any manifestation of life and even that life necessarily appears in matter thus elaborated, it does not follow that the result of the processes by which matter is fitted to receive life is the *cause* of the vitality manifested. For all that is known to the contrary protoplasm and vitality are simply concomitant; or if there is any causal relation between them, vital force is the cause of the peculiar properties of protoplasm, not the result of those properties. There really exists a potency or principle called 'vital', in virtue of which the chemical substance called protoplasm manifests vitality, that is to say *is alive,* and in the absence of which no protoplasmic or other molecular aggregation of matter can be alive. The chemico-physical theory simply restates abiogenesis, or 'spontaneous generation', of which we know nothing scientifically. The grave doubt that 'life is a property of protoplasm' will persistently intrude until some one shows what is the chemico-physical difference between living and dead protoplasm; none being known. The speaker argued for the existence of the soul as something apart from and unlike matter, defining 'soul' as that quality of spirit which any living body may or does possess. No idea can attach to the term 'spirit', from which all conceptions of matter are not absolutely excluded. Spirit is immaterial self-conscious force; life consists in the animation of matter by spirit.

The substance of mind and the substance of matter were noted as equally hypothetical. To the former was given the name *Biogen,* or soul-stuff and it is defined as spirit in combination with the minimum of mater necessary to its manifestation. The analogy between biogen and luminiferous aether, or the hypothetical substance of light, was discussed. The drift of the speaker's speculation on the vital principle as an *ens realissimum* was toward a restatement, in scientific terms, of the old *anima mundi* theory. Modern materialistic and atheistic notions about life were denounced as every one of them disguises of the monstrous absurd statement that a self-created atom of matter could lay an egg that would hatch.

The whole matter being beyond the scrutiny of the physical senses is remote from the scope of exact science; but it is irrational and scientific to deny it, as is virtually done when science excludes it from any share in life-phenomena, by presuming to explain life upon purely material considerations. No chemico-physical theory of life is tenable which does not satisfactorily explain the difference between, for example, a live amoeba and a dead one; an explanation which has never yet been, and probably cannot be, given.

A general discussion of the points involved in this paper followed. Mr. Powell pointed out what he regarded as fundamental and fatal error in the reasoning, viz., that the axiom that the whole equals the sum of all its parts, had been assumed throughout to be true *qualitatively* as well as quantitatively. Furthermore, he maintained that logical consistency required that those who believed in force should also believe in the vital principle, and *vice versa.* As for myself, however, there was neither force nor vital principle, but only matter in motion. Three relations are always to be borne in mind, viz., quantity, quality, and succession, whereas the physical falls into error by considering only the quantitative relation.

So much of the support of the views of Mr. Coues as might be derived from the common consensus of mankind was criticized by Mr. Gill as unsound, since the common consensus of mankind has often been found at fault; the supposed flatness of the earth, the motion of the sun around the earth, etc., are examples where this criterion fails. Paraphrasing an eminent philosopher's dictum, he thought there was a tendency in biologists ignorant of philosophy and philosophers ignorant of biology to make a distinction between organic and inorganic matter, and call in a 'vital force'. He likened living and dead protoplasm to an electric battery in action and at rest, and maintained that life is a property of matter, and that it cannot be conceived of separated from matter.

Mr. Harkness avowed his belief in force, and hence in vital force, and further in a little religion, and was, therefore, moved to make inquiry concerning the chemical difference between living and dead matter.

Mr. Ward pointed out that very diverse views were held upon this subject by two classes of thinkers who do not come into intellectual contact. Furthermore, while not asserting that vital force was a superstition, attention was drawn to the fact that infantile races attribute all phenomena to supernatural agencies, and that, with increasing knowledge, there is a decrease in the number of these appeals to supernatural agencies. "The cornerstone of modern science," said Mr. Doolittle, "*measure*. We must have a biometer. What electrical science would be without ohms, astronomy without graduated circles, chemistry without balance, such is biology without a *measure*. Is there more life in two mice than in one mouse? In a horse than in a mouse? Until we can answer these questions substantial progress in biology is not to be expected."

After some further desultory discussion the Society adjourned.

Biogen

Mr. President and Gentlemen of the Society:

Any reason I might have found for declining your invitation to speak on this subject could only have come from moral cowardice. I should have had, therefore, no alternative to compliance, even had I had no courage to proceed but that of conviction. But I was given to understand that you might neither be unwilling to have the general biological problem reopened, nor indisposed to hear with forbearance at least from anyone of your number who might have ideas upon the subject, with a view to discuss such propositions as he might be willing and able to advance.

So far am I from supposing that the *crux* of the life-problem will be solved tonight, and that it is insoluble in any royal water that can be compounded by to-day's science and philosophy. Confronted as I am with something I believe to be inscrutable to man's unaided reason – opposed as are my convictions to some of the brave theories which have been advanced in the Society respecting that something – profoundly unknowing as I am of the origin and nature of Life, I should desist from this honorable confession of ignorance and seek its asylum, were I not also convinced that much truth in the matter of the life-problem is to be had for the asking by anyone who makes full use of *all* his faculties; were not my views in the main those which, in substance, under whatever form of expression, have been affirmed by the consensus of mankind since when the human creature became possessed of a rational soul; and were I not satisfied that anything I could say, seeming new and being true, would be no news, but something as old as the mind of man.

In expressing one's self upon matters which are rather those of reasonable inference than of demonstration, there is danger of dogmatizing just in proportion to strength of the belief; but that unscientific, unphilosophical, and offensive practice is avoided when individual convictions are given with the reasons upon which they are based, with perfect intellectual candor, with deserved contempt for mere logomachy, and with due deference to those different opinions which may be but varied views of a single many-sided truth. In such spirit as this, I beg your indulgence in a train of thought not put together to sustain any theory of my own, but to discover truth, if possible.

It cannot be amiss to bring up certain papers lately laid before the Society, and treat them as if under discussion to-night. One of these is Dr. J.J. Woodward's address, as the retiring President, on "Modern Philosophic Conceptions of Life," and others are Mr. Lester F. Ward's, on "Evolution of the Chemical Elements" and of the "Organic Compounds". If I correctly appreciate their respective significance, they embody opposite and probably irreconcilable views – Mr. Ward setting forth the chemico-physical theory of life and Mr. Woodward inclining to what may be termed the vitalistic theory. [1]

1

The speaker quoted as follows from Dr. Woodward's published address:-

p. 18. "I have already asserted that there are whole groups of phenomena characteristic of living beings and peculiar to them, which cannot be intellectually explained as the mere resultants of the operation of

As a master of many departments of science, and in a masterly manner, Dr. Woodward appears to have reviewed much that is actually known of the conditions and manifestations of life, with a fair statement of much that is unknown, arguing against the adequacy of the chemico-physical theory, maintaining the existence and operation of a "vital principle", and declaring that while the idea of a universal creative mind has claims to be tenable scientific hypothesis, neither science nor philosophy

---

the chemical and physical forces of the universe. These phenomena I refer – I own it without hesitation- to the operations of a vital principle, in the existence of which I believe as firmly as I do in the existence of force, though I do not know its nature any more than I know the nature of force."

p. 20. "I willingly admit that, in view of our present notions of the cosmogony, it is impossible to believe that life always existed upon this planet. I willingly admit that life on the earth must have had a beginning in time. But we do not know how it began. Let us honestly confess our ignorance. I declare to you I think the old Hebrew belief, that life began by a creative act of the Universal Mind, has quite as good claims to be regarded a scientific hypothesis, as the speculation that inorganic matter ever became living by virtue of its own forces merely."

P. 20 There is ..." a philosophy which recognizes the validity of the mind's self-consciousness as at least fully equal to the validity of its consciousness of the condition of the body by which it obtains a knowledge of the external world. By this self-consciousness I now, with a certainty which no doubt can ever disturb, that I have a mind.; and by rightly applying my reasoning power to the data of my self-consciousness I can learn much that will be useful to me with regard to my mental processes and the methods of applying them. But here I have to stop. I can learn nothing, whether by consciousness or by reasoning, with regard to the real nature of my conscious mind, and however much it may long for immortality, neither philosophy nor science afford any foundation of proof upon which I may rest."

affords any prove basis for the most universal of human beliefs – the existence in man of an immortal soul. Passages that have been quoted show their author to be satisfied of the insufficiency of science and philosophy to explain the mystery of life, and so explain himself to himself; so what if he desires that which most men desire, he must look elsewhere for the satisfaction of that desire. I doubt not most honest thinkers have found precisely the same difficulty. It is a very grave one, which usually increases, instead of diminishing, the farther we go into the curriculum of the natural sciences in our reliance upon "pure reason" – a lamp which finally serves not to light the way, but only to make the darkness visible.

     I refer in the sequel to what I understand the "vital principle" to be. But first to touch upon the "chemico-physical theory of life", as maintained by Mr. Ward, who needs no reassurance of the profound respect I have for his intellectual processes, widely as we differ respecting the validity of his results; whose logic is so clear and cogent, whose illustration is so copious, that his conclusions would be inevitable were his postulates admissible. The flaw seems to be in the indictment by which matter may literally said to be put on trial for its life. The central idea of his papers on the evolution from nebulous matter of the chemical elements and all other known forms is, - progressive increase in complexity of the molecular units of all substances, with corresponding increment of molecular mass and corresponding decrement of stability of combination[2] - such molecular aggregates progressing in instability until a stage is reached where the resulting aggregations, no longer molecular, but rather molar, are so unstable that new and higher activities become possible, and perceptible molar movements may take place; these actually occurring at the stage of aggregation reached by the substance called "protoplasm"; life consisting in such mode of motion as the particles of protoplasm manifest, and being therefore a property of protoplasm, an essential or intrinsic quality of matter, in virtue of its own mechanical and chemical forces; in other words, that life inheres in matter, and is simply the resultant of material forces; "the most profound truth, both biology and chemistry," being , in Dr. Ward's view,

"that life is the result of the aggregation of matter." [3]

---

[2] The expression of this idea ascribed to Socrates by Plato is, - that compounding things, or such as are compoundable, *admit of being dissipated at the same rate that they were compounded*

[3] The proposition above stated are summed in their author's own words as follows: -
"The general law above stated, that in the process of the evolution of matter from the simplest elemental state to the most complex organic compound, there has constantly been increase in the mass and decrease in the stability of molecules, holds good throughout; and to it may now be added a third principle, obviously correlated to the above and constituted merely a corollary to it, that *pari passu* with these changes there has been an increase in the activity of the properties manifested....In protein bodies these molecular activities are much more extensive and varied than are those of simpler bodies. The molecular units are so much larger that their motions must be, as it were, *molar* in

Biogen

I have never seen elsewhere so fair a statement
of the chemicophysical theory, so ably supported; and
the chain reasoning by which diffuse nebulous matter
is linked to the tissue of living things appeals to my
mind with great cogency. But I think the lurking fallacy
is no less dangerous than deplorable.
For, granted that all substances, including
protoplasm, have been evolving from nebulous matter;
granted, that evolution to the protoplasmic state, and
in the very manner claimed, is required for any
manifestation of life; granted even, that life always
appears in matter thus elaborated; it does not follow,
that the result of the process by which matter is fitted
to receive life is the *cause* of the vitality it manifests.
Sequence is not necessarily consequence; and in this
case it does not seem that even a *post hoc, much less
a propter hoc*, can be maintained. For all that is known
to the contrary, protoplasm and vitality are simply

---

comparison, while within these larger primary units there are lesser units
of different orders of aggregation, each of which manifests its own
appropriate activities, and thus modifies the general properties of the
whole....From the molecule of hydrogen to that of albumen the process
of evolution has been uniformly the same, namely, that of compounding
compounding and recompounding, of doubling and multiply: in short, it
has been the process of molecular aggregation. It would be contrary to
the law of uniformity in natural phenomena, upon the recognition of
which modern science is based, to assume an abrupt change in the
process at this point; and upon those who maintain such a *saltus* must
rest the burden of proof....That the recompounding of the protein bodies
should result in a new form of processing the quality of spontaneous
movement is a *priori* just as probable as that the addition of a molecule
of oxygen should convert the hybrids into alcohols."

concomitant. If any causal relation is to be established, it must be upon other considerations than have been presented. I believe the relation to be causal, but the reverse of that claimed; vital force being the cause of the particular properties of protoplasm.

I adhere without reservation to the doctrine that there really exists a potency or principle called "vital", in virtue of which the chemical substance called protoplasm manifests the rudimentary phenomena of life; that is to say, *is alive*; and in the absence of which no protoplasmic or other molecular aggregation of matter can manifest such phenomena; that is to say, *be alive.* Chief among the impossibilities of protoplasm appears to me to be the spontaneous generation of life by any method of chemical or mechanical movement impressed upon matter by the operation of forces inherent in itself. That the chemico-physical theory is merely a restatement of the theory of "spontaneous generation" is self-evident, and the difficulty is increased by the assumption that mechanical and chemical compoundings of matter are adequate to result in life. It is an unquestionable scientific fact that spontaneous generation has never been demonstrated to have occurred in a single instance, with or without operation of vital force in addition to purely physical forces, though every supposed condition of vitality has been artificially brought about. The scientific fact is- and by scientific fact I mean something positively known to be true-that life has never been ascertained to have any other origin than antecedent life. For all that is known to the contrary, such antecedent is no less necessary to the existence of vitality than is protoplasmic matter necessary to the manifestation of vitality. The grave doubt that "Life is a property of protoplasm," results from the way in which the particles of that substance are aggregated and arranged, will persist obtrusively, I think, until the chemico-physical theory accounts for the difference between a live amoeba and a dead amoeba. I should say there is all the difference in the world, and that this difference is just the point at issue. Until that explanation is forthcoming, the theory mentioned remains not a logical inference, but a pure assumption –a hypothetical link in the chain of being found just too short by one link.

I recognize the fact, which no biologist questions, that life may and does precede "organization", and therefore exists in matter independently of organization. Since an amoeba exhibits the rudiments of organization, having a nucleus and often a membrane in addition to its substance proper, let us take a still simpler living thing- a plasson body [4], which is merely a particle of animated matter, shapeless, structureless, unorganized, and absolutely homogeneous; yet manifesting, for an allotted period, the phenomena necessary to any predication of life, namely: it moves, it feels, it propagates, it may be killed; and these things could not be were it not alive. The physical properties of the plasson-body, which is simply unorganized protoplasm, are well known to us. Its chemical composition, as given on good authority, is, in 100 parts, 54 of carbon, 21 of oxygen, 16 of nitrogen, 7 of hydrogen, and 2 of sulfur. But, has a

---

[4] Plasson body, plasmosome, biophore, one of the hypothetical fundamental units of germ plasm.

living plasson body ever been resolved into its chemical elements? I should think it would be thoroughly killed before the analysis were over. If so, living protoplasm has never been and cannot be analyzed, and its chemical composition remains unknown. For, according to the chemico-physical theory, it lives only in virtue of its peculiar chemical and physical constitution; it lives necessarily, simply because it *is* protoplasm; but, if so, protoplasm is only itself when it is living; when it is dead, it is something else; therefore, this something else is what is analyzed; and in what life is has eluded the process. A contradiction in premises here implied, and an absurdity is made manifest; for if there be any knowable difference in chemical and physical constitution between a living and a dead cell, or other protoplasmic body, such difference is unknown; to all physical tests that have been applied, they are identical. I anticipate the ready reply, that chemistry only claims to know what elementary substances, in what proportion, constitutes protoplasm, not pretending to say what particular manner of aggregation of their molecular units results in life. But such answer, so far from doing away with a physical difficulty, seeks refuge in a metaphysical subtlety. For if life necessarily resulted from the compounding of certain elementary substances in certain proportions, and in a certain way, there is present and operative *something* adequate to affect such result, absence or non-operation of which something results in death. Because, the moment these identical elementary substances, combined in the identical proportions, slip into any other molecular interaction or molecular inter-adjustment, they cease to manifest the phenomena of life. What holds them just as they are in life, neither chemistry or physics shows. I give reason, beyond, for assuming that the *something* is that peculiar thing called vital force. This hypothesis is *a priori* as legitimate and reasonable as any other can be in the case where all is as pure speculative as any metaphysical question can be. For all that relates to the ultimate atoms of matter – supposing any such things to exist – to their number, size, shape, mass, distance apart, mode of motion, and interaction, is beyond human scrutiny, and, therefore, remote from the domain of exact science.

If such considerations have any weight, the theory under discussion would appear to proceed in a logical manner from purely speculative premises to an equally satisfactory conclusion. It is not on scientific ground until it explains what physical and chemical differences there is between a living and dead plasson body; for the difference must be physical and chemical only, since only physical and chemical forces are admitted to be concerned in its production. Chemistry and physics finding no difference, we may be permitted, indeed we are obliged, to look elsewhere for explanation of the very great difference obvious between a thing alive and the same thing dead.

Numberless organic compounds have been manufactured in the laboratory which differ in no wise from the same compounds affected in nature by vital forces excepting that they have never shown a trace of life; so that I should say that the absence or presence of that essence is precisely the difference between the artificial and the natural product. In short, physics and chemistry have combined to manufacture an egg which will do everything you could expect of an egg, except hatching. Pardon me if I go a step further, and sum the charge thus:

The atheistic physicist, denying mind in nature, declares that matter alone exists. Where it came from is no matter. It exists; it is matter in motion. Matter in motion is all there is in the universe. The cosmos is matter in motion, in virtue of its material forces alone.[5]

But does it occur to such a physicist that he has invented just what he has always declared to be a physical impossibility? For he has simply invented a huge "perpetual motion" machine, which runs of itself until it wears out or breaks down. Worse than this, he literally forgets himself, the inventor, for he says his machine invented itself and set itself a-going. Then the materialistic chemist takes this self-invented perpetual motion machine invented itself and set

---

[5] "Give to the ambitious kinematic artist his cloud of sand, - or if he prefer the outfit, let him be furnished with an indefinite quantity of a perfectly continuous incompressible fluid – bound up if you please in a chain of 'vortex rings, - by no motions or compositions of motions – continued through the aeons of eternity – could he ever manufacture there from either a lever or a rope. The kinematic gospel of a mechanical theory of primeval motion is therefore a sophism and illusion. It is founded on a misconception of the very essence of true mechanics. And the system that would proudly aspire to an architecture of a Kosmos from the elements of matter disrobed and denuded of every quality but motion, could achieve as its highest triumph and product – a universe of dust and ashes."
Taylor, Bull. Philos. Soc. Washington, v. p. 167

itself a-going. Then the materialistic chemist takes this self-invented perpetual motion machine and declares that it has laid an egg that will hatch.

Thus far we have only stood on the threshold of life, to witness such faint beginnings of vitality as a speck of protoplasm exhibits. On any theory that the physical forces inherent in matter alone are concerned, the way darkens as we proceed from moner to man. Few persons are more thorough and consistent Evolutionists than I may claim to be, and if you give me a live plasson-body I will engage to make a living human body out of it on the most approved biological principles. In fact, we know that the physical bodies of all organized beings consist either of a single cell or of a multitude of cells, each of which is, in effect, and individual plasson-body, born of a parent like itself, living for a while in the enjoyment of its appropriate activities, and then dying. The human body consists of a myriad such plasson-bodies, not all alike, indeed, but become very different in form and function in their descent with modification from their common progenitors, the female ovum and the male spermatozoon – their differentiation of structure and specialization of functions of the various tissues of the body being such that the result may be aptly compared to a society of different species of amoeba like animals, - bone-amoebas, brain-amoebas, muscle-amoebas, and the rest; all the individuals of which species of animals are in ceaseless process of birth, growth, maturation, decay, and death. Such language is, of course, not figurative illustration of an idea, but simple statement of observed fact. I am ready to believe, and I do, that the chain of life is unbroken from moner to man, missing links being only hidden links, so far as the genetic relation of the physical body of a man to the same of a moner is concerned. But now I find myself not only tossed on the horns of a dilemma, but lost in the intricacies of a polylemma, to extricate myself from which all the natural potencies to be found in the physics and chemistry of matter have, in fact, proven their inadequacy.

First, if the chain of living being has a beginning and an end, anywhere, anyhow, at any time, the links overhauled fall short in both directions. For, at one end, the original arch-amoeba as much a mystery as ever; we know not where he came from, how he got there, and in what the essence of the essence of his plassonity subsists. At the other end, we find our bodies to be a menagerie of amoebas, which we cannot dispose of intelligently, and the finale of which is as much a mystery as their origination; seeing that we know not what, if anything, will happen when our death disperses them.

Second, if the chain of living being is endless, it necessarily returns upon itself, and all reasoning upon its course is reasoning in a circle. We simply say that if A is B, B is A; which proves nothing as to the nature of A or B.

Thirdly, no whole can be greater or less than the sum of its parts, or quantitatively different from such sum. But a particle of living plasson is greater than the sum of its known parts, possessing that which none of its protoplasmic parts possess – Life. And, *a fortiori* the highest and most complex organism, man, possesses many things than none of its protoplasmic parts possess, unless such things as will, memory, and understanding – such things as faith, hope, and conscience, are properties of protoplasm; it being indisputable that such qualities and attributes do reside in human beings, if our consciousness and our senses have any reliability; and if they have not, we know nothing whatever.

Once more, and especially, if the universe is a self-invented perpetual-motion machine – if matter has always and alone existed, and has always had the self-determining potency of life, and at length did so determine itself to become living, and if man, the final outcome, is self-determined protoplasmic material only, a God is not only superfluous but impossible. Yet the result of the alleged self-evolution of self-created primordial matter through chemical elements to organic compounds has been the creation of a protoplasmic mind, so constituted that in the overwhelming majority of instances it can and does, and must, believe in a God. If matter be that God, matter contradicts itself, for the constitution of the human soul requires that its God shall be other than its protoplasmic self. If matter be not that God, there must be some other. A protoplasmic mind can only escape that conviction by denying that itself exists; which would be absurd, were it not impossible.

The evolution of human reason and human faith, in short, of a "rational soul", being among the possibilities ascribed to protoplasm, or some ulterior compounding of matter, a train of consequences ought logically to follow, which, in point of fact, do not follow. The almost universal sentiment of mankind is religious in some kind or degree, and certain aspirations are the common endowment of our race. Those whose Deity is protoplasmic probably never worship that substance, and in fact appear quite indifferent to its divinely transcendent attributes; yet some form of worship is omniprevalent; and if protoplasm be not a proper object of worship as a creative omnipotence, and be not capable of satisfying the aspirations it has evoked, it is fallacious and delusive, through failure to fulfill its own conditioning of human reason and the faith of mankind. In a word, it gives the lie to its own logic.

How it may appear to others, I, of course, cannot say; appeal to the data of my own consciousness decides the real gravamen of the objections I have raised. Argument is futile. I can only declare that I do not believe my mind to be matter-made only, because it is so made that I cannot so believe, seeing not the slightest reason therefore. If I be wrong, it is some consolation to reflect, that so far from my being peculiarly deceived, the consensus of mankind has reached the identical conclusion; so that any required asylum of ignorance proves to be the common refuge of humanity. Nevertheless, such views as these, however useful or even precious to myself, remain mere professions of faith, of little or no consequence to others, until reasons are adduced in their support; and iconoclasm has but its trouble for its pains, if it replaces no broken images. I think it will be conceded that all the conceptions of mind which have swayed the scientific and philosophical minds of men, are more or less hypothetical, and in their essence purely speculative. This seems necessary when, in the nature of the case, no theorem is demonstrable, and degrees of reasonable probability are the uttermost approaches to the heart of this fascinating inscrutability, respecting which the *credo ut intelligam* of the theologian complements the *cogito ergo sum* of the metaphysician; belief being no less postulated by reason than is being affirmed in thinking.

Such apology, if any is needed, is all I have to offer in opposing the spontaneous generation speculation by the vitalistic theory, and proposing to recognize the hypothesis of the God-made cosmos, instead of the hypothesis of the self-made perpetual-motion machine.[6]

Life in the concrete is, of course, the sum of the phenomena manifested by nature. Of life in the abstract, of the essence or nature of that peculiar attribute of plants and animals, apart from its material manifestations, no knowledge whatever seems possible. Yet, while I cannot even imagine what life is or may be, apart from matter, so far is it from being possible for

---

[6]

"This ultimate and highest induction of scientific thought - the Inscrutable made Absolute – is restful and satisfying. This ultimate and highest induction – as highest and ultimate – cannot be manifested as a 'working hypothesis'. His ultimate and highest induction – as such- cannot be subjected to the subsequent verification of mathematical deduction. This ultimate and highest induction detracts nothing from the certainty of orderly sequence so irrefutably impressed upon us by every deepening channel of research, but gives us rational ground and guarantee of such unfailing regularity. This ultimate and highest induction, accepting to the uttermost the material interpretation of nature's administration, - whose ceaseless Evolution seems ever opening up new vistas of automatic teleology, - gives significance to our imperfect conception of a regulated system, (so necessarily involved in the very existence and operation of a 'machine') and accounts consistently for the unfaltering obedience and instantaneous response of all the countless atoms of the universe to the rein of 'law', by positing behind such law – an Infinite LAW-GIVER." – Taylor, loc.cit., p. 173

me to conceive of life as an existent reality apart from any *known conditions* of matter, that it is impossible for me not to form that conception. This is of course to invoke the "vital principle", to postulate the reality of a kind of force called "vital", as a veritable Biogen or life giver, which may be were no known form of matter is, and can, therefore, exist apart from such matter, and not as a resultant of any material forces. Though this is pure speculation, I am forced so to speculated, in the impossibility of conceiving the contrary. The conception does not imply that vital force differs from other forms of cosmic energy otherwise as different branches form one stream; for all force is one, however diverse its ulterior operation; the kind of force called "vital" being that special potency under the agency of which matter assumes the form and function of life in the concrete. Force cannot act where it is not; neither can it act with nothing to act upon; its presence in and operation upon matter are, therefore, necessary conditions of its manifestation; all the manifestations of life are ultimately resolvable into modes of motion, and in the particular modes of motion exhibited by living things, *and by no others,* are evidenced the presence and operation of the vital principle, the energy of which differs from other energies precisely as the modes of motion of living things differ from those of all things that do not live. This is not a verbal distinction merely; if it seems so, the fault is in the obscurity of my expression of the perfectly clear idea everyone has of a difference between that which is alive and that which is not. It subsists in the presence or absence of something – some real entity, which defies observation by the senses and, therefore, cannot be described; but the results of which are exhibited most unequivocal manner. If pressed for more concise statement, I may turn the expression, say that life, so far from being the *result* of the aggregation of matter, in consequence of any conditionings known as chemical or mechanical, exists apart from matter, as a *vera causa*, preceding the organization of matter; life being, in short, the *cause, and not the consequence*, of organization. It certainly precedes organization and exists in unorganized matter, as any scrap of living plasson demonstrates. Furthermore, the highest known grade of organization, as the body of a man, though never attained except through vital force, may and does exist without life, as any corpse mutely testifies until decomposition or disorganization sets in. If life inhered in matter, death would follow decomposition, and be otherwise impossible; but I fact the reverse is the actual sequence of events.

If there be any truth in the statement that life is an entity, a reality, apart from any known forms of matter, it perfectly logical to speak of its presence in or absence from any given mass of matter; and this was my idea when noted the sum of a living being as greater than the sum of its dead material parts. I also used the word "God" when satirizing the apotheosis of protoplasm. I have thus far purposefully restrained from using the word "spirit". But I cannot precede with my idea of life without introducing that term, to which I am aware much of the accredited science and philosophy of the day objects, as being "found without sense". Yet no scientist who acknowledges the validity of the science of psychology, and no philosopher who recognizes the validity of abstract ideas, objects to the word "mind". I must therefore be permitted to speak of spirit, or "soul", if you please, as something which, like mind, is a legitimate subject of inquiry: first, as to whether it exits or not; second, if it exist, whether it be of protoplasmic nature or no; third, if it be not that product of the aggregation of matter, what sort of a product it might be; for I consider this inquiry especially pertinent to any discussion of life. Our alternative, you know, is, that all vital phenomena, all manifestations whatsoever of life, are to be counted among the accomplishments of protoplasm, or are to be otherwise accounted for.

Much difference of opinion as to the reality of "soul" might be reconciled if disputants could catch each other's meaning and agree upon a definition of the term. But this is very difficult, though we all think we know what is meant when a soul is in mention. Many deny there to be any such thing; many waive the question, neither affirming nor denying; most ascribe a soul to man alone; some concede a soul to every atom of organic matter as well as to all organized bodies. My view defines soul as the quantity of "spirit" which any living being may or does possess at any time. But this requires a definition of "spirit", some quantity of which is to make a soul, just as some amount of matter makes a body. I can attach no idea to the term "spirit" from which all conceptions of matter are not absolutely excluded. Spirit is nothing if not immaterial. Force is likewise immaterial; but I think nearly all persons recognize a distinction between spirit and any mechanical force, such as gravitation. My mind affords no distinction between spirit in its totality and that Universal Mind or Supreme Intelligence, which we mean when we speak or think of God.

To my mind, "mind in nature" is a self-evident proposition - a logical necessity. The simple fact that we can *think* a God, necessitates the conclusion most men have reached, of the existence in nature of other than what are called "natural forces"; of the reality of the existence of spirit as self-conscious force; though I do not see why it is not as "natural" a force as gravitation. It is certainly not unnatural; and to call it "supernatural" only exposed our ignorance of Nature – Nature being, on any theistic hypothesis, simply the sum of the manifestations of the will of the French epigram, "If there be no God man must invent one," may be paraphrased to say, "If there were no God man could not invent one." I cannot suppose my mind to be particularly constituted; and, if I find the conceptions just noted present in it, as propositions that are nothing if not self-evident, if not axiomatic data of consciousness, I presume the same idea can or does present itself to most other persons. But by our definition, "soul" is a portion of spirit, and spirit is self-conscious. I am likewise self-conscious; and by that quality of being I know, with a certainty no doubt can disturb, with a certitude no argument can increase or diminish, that I have a soul. For to doubt is to judge; to judge is to reason; while the knowledge I have of my own soul comes not by taking thought; it is the soul's self-consciousness. Some call it "faith"; I have no objection to that term; it is something so precious, so superior to reason, though never irrational, that I would greatly prefer to recognize it as a property of protoplasm than to lose it.

Finding myself also in possession of a body, of the actual existence of which body few persons, except some German metaphysicians and their suckling converts, are in doubt, and also observing that this body is alive, that is to say, that it manifests all the phenomena necessary to our conceptions of life, I am bound to infer, and I do infer, that in my own case at least, life subsists in the union of soul and body; that life consists in the animation of matter by spirit; that life is God made consciously manifest. If there be any truth in this, I suppose it is equally true of other human beings, though I only answer for myself.

My mind refuses to believe, what some may object, that such expressions as I have used respecting the reality of spirit are mere abstractions – mere metaphysical subtleties metaphorically expressed – in other words, mere figments of the imagination. I would sooner grant, what some metaphysically fancy the have proved, namely, that we have no bodies. To do away with the body altogether – at any rate, which every body excepting one's own, appears to be one of the accomplishments of some schools of thought. Some exploiting in the saw-dust of an intellectual gymnasium seems to me a simple and easy trick, in comparison with the effort to deny the soul; for the body is but an accident of matter, and the process of annihilating it in imagination only anticipates a natural process by a brief span of time, and time is nothing but a sequence of events which cannot occur if there is nothing to happen.. But to do away with spirit, even in imagination, is not naturally possible. It is futile to attempt, as some "philosophers" have done, to avoid all possible contradiction, and evade the possibility that "pure reason" may be fallible or fallacious, by denying the existence of the subject of every possible predication, thus evolving a "philosophy" of which universal negation is the sole final outcome. What philosophy – what "love of wisdom" – is here, when we are left nothing to love! For the act of denial, or even refusal to affirm, implies a denier or a refuser, as an existent reality, and the denial or refusal is itself an existent reality. True it is, and dismally true, that the philosophy of universal negation, by some called "criticism of pure reason", is intellectual nihilism – a sort of philosophical fool's paradise, or earthly Nirvana, where one has not even the Buddhist privilege of mumbling "Aum", for fear there may be some mistake about it. Such a state of mind is not even an asylum of ignorance in which poor humanity may take refuge; it is an asylum in which intellectually impotency holds not the mirror up to nature but to its confessed self.

But this is unpremeditated digression. The point I wished to make, when those contemptuous thought obtruded, is, that a denial implies a denier, and as both are real entities, though denial is an absolute immateriality, the real entity of such an equally absolute immateriality as I hold spirit to be is not *a priori* impossible. It *may* exist therefore; I have postulated that it actually does exist, and defined it as self-conscious force; I have speculated that a living body results from the action of spirit on matter, and that life subsists of the union of the two. To bring the question to some scientific shape – to put it on the border-land between metaphysics and psychology, it not really in the domain of the latter science, let me say a few words respecting the connection between mind and matter.

The only points toward which all difference of opinion in this vexing question converge are the intimacy of the connection and the intricacy of the relation in which the two factors – mind and matter – are inter-dependent and inter-active. For even those who hold, as I do, that mind does not depend upon matter for its existence, but only for its manifestations, if they know anything of anatomy and physiology, know how powerfully physical states affect mental operations. Those who maintain the chemico-physical theory of life necessarily consider all mental life, like all physical phenomena, as the resultants of the play of mechanical forces, and as ultimately referable to mere motion of material particles- such mental endowments as will, memory and understanding, judgment, intuition, perception, conception, conscience, and consciousness itself depending for their existence upon how stands the parallelogram of forces, how goes the balance of power in the mad clash of blind atoms. My hypothesis, which recognizes the existence of spirit as determining life, and makes life the cause instead of the consequence of organization, enables us to reconstruct the parallelogram of forces, and strike the balance of power not between the mechanical forces of the material particles themselves, but between these and the conscious power of spirit – the Will of the Ego. This is the resultant which apparently constitutes "mind". Viewing the intense and vivid molecular activity, the combustion and deflagration of tissue, which attend the generation of every thought, and are necessary to the manifestation of thought, though in no sense its originator, it is scarcely using metaphorical language to say that mind resides at the melting point of matter in spirit. [7]

7

See appendix, 6th paragraph. In penning "Biogen" for oral delivery, I purposely followed that common usage of the word "spirit" and "soul" which makes these two terms synonymous, or at any rate alternative, expressions for all that there is of man

To illustrate such fusion as I have imagined, let us consider the two opposite things which, according to universal experience, concur in the alembic mind. I refer, of course, to any subjective and any objective cognition. Let us formulate any subjective cognition in the general expression "I will", and any objective cognition in the term "I see".

Aside from the summary cognition "I am" nothing can be conceived more original spontaneous, independent, and self-determining than "I will". This cognition affected, at whatever expense of brain tissue, *will power* has been consciously called into being; it has been created; it exists as a real entity, at the service of its originator, to be used as he determines. This seems to be the purest example of *force* of which any one can be conscious. To think "I will" is to command force. But so long as this conscious determination remains inoperative, it is only potential energy or latent force, which may or may not become active and effectual. If it so not act effectively upon *something,* no manifestation of power is possible, and the very existence of the energy is unknowable, excepting to its creator; it is only self-existent, in short. *Once translated in terms of matter,* with motion or any other cognizable effect, the existence, operation, and result of a cause are discovered. If we knew how this

---

which may survive the death of the body; not desiring to open any discussion of the point involved here. The distinction I make is formulated and definitely set forth in the Appendix, where also will be found some further reflections upon the meaning of the word "mind" – mind not being a thing which thinks (for that would be spirit) but the expression of what is thought.

translation is accomplished, we should know exactly how the connection between mind and matter is made; but we do not, and can only rest in the knowledge that somehow the brain is a material mechanism by which the will of the owner of that apparatus is primarily manifested. Will-power is carried out further by the rest of the bodily machinery, and may be finally accomplished in a thousand ways. But observe, that all such manifestation of force is the manifestation not of mechanical or chemical force merely, but of *intelligent volition*; that is to say, of self-conscious force, which, according to our definition of spirit, is spirit.

To many minds it might be to sow the seeds of reverence for the exalted dignity of humanity to reflect that such mental operation as I have described is the counterfeit, in the finite human microcosm, of the described creation of the macrocosm by infinite power divine. The Universal Mind, the Supreme Intelligence, the great I Am, which was and is and shall be always, determined, it is said, to become manifest. He said "let it be", and there it was, as He willed. And man is said to be made in His image.

Now let us glance at the other chain of sequence – that involved in the term "I see", as the general expression for all sense-concepts. One would think this a very simple proposition; so it is, if its full meaning is but half grasped, or even only just missed, the proposition is unintelligible. Such appear to be the difficulty with those who, for the simple truth "I see", try to substitute the untruth, "The brain sees"; for they fail to see at all through the mass of squirming brain-amoebas which are tormented to death in the process of their reflections on the subject. No one supposes the eye sees, any more than any other optical instrument sees; nor the optic nerve any more than the eyeball; nor the corpus quadrigemina than the nerve; yet there is a blind kind of physiology which seems to think that vision, the faculty of seeing, which cannot be found at the other end of the optical instrument, must lurk about the inner end of that exquisite apparatus. But I must believe, as I do, that, trace the nerve threads as far back as you please, and locate the exact spot in the brain where they end, there would be no seeing if some Ego – that identical Ego I postulate – were not looking through he telescope life has organized for the purpose, and is fully conscious of seeing, with the same certitude that I know who is speaking; and I do *not* believe anyone of you to be differently constituted in this respect. Truly, the difficulty of understanding *how* the physical terms of a retinal image can be translated into the mental terms of conscious vision, has never been overcome; our ignorance is absolute; if it ever is overcome, no doubt we shall learn what and where is the connection between mind and matter.

I speculate that not only is it among the possibilities of living protoplasm to establish that connection, but that among the qualities of that pregnant substance, or of some of its material derivatives, is one adequate to the establishment of the required relation. Chemistry has shown the composition of the dead substance – the number and proportions of the elements compounding it – even the mode in which its molecular units are, or may reasonably be inferred to be, compounded. The extreme instability of the resulting combination, and the extraordinary activities acquired, are well known. If we can be permitted to vivify such a dead substance as this with biogen or any thing else, it is difficult to set any bounds to its possibilities as a mediator or go between mind and matter – in short, between the spirit and the body. I hypothecate for living protoplasm – for the dead substance the chemist knows *plus* biogen, a vastly greater degree of molecular instability, and immeasurably more energetic molecular or perhaps atomic activity, than have been ascribed to the dead tissue, simply as an extension of the conditions which have been ascribed to dead protoplasm as laws of its chemico-physical being. I speculate upon the reasonable probability that under the influence of vital force protoplasm may and does acquire such tenuity of substance, such mobility and activity, as to be fairly describable as matter at a minimum of density combined with force at a maximum of intensity; and to be comparable in such vital stage of its evolution to that interstellar fluid which is scientifically recognized as the medium of the transfer of force elsewhere. If the undulations of a luminiferous ether – a substance vastly more tenuous than any we know by our senses, yet substantial still, and perhaps still far from the dividing line between matter and spirit, where pure spirit is purged of the last dregs of materiality – if such an aether, the very existence of which is hypothetical, yet an accepted scientific fact, because no other effort of the imagination supplies so good an hypothesis on which to explain the phenomenon of light – if this aether can be logically inferred to exist, it is no romance of the imagination to infer that matter may be animated to the degree of sublimation required for its vibration to will-power –its thrilling to a thought.

Such state of matter as I imagine and describe would satisfy at least one of the important factors of the life-problem, by establishing a connection between mind and matter. The thing is already done when a single atom of matter is moved in the least by the slightest conscious force.

I have often thought that the phenomena of life may be instructively compared with those of light, there being some highly suggestive parallelisms between the two things. Life and light are curiously coupled in vulgar parlance, an unsophisticated mind vaguely perceived some similarity, just as it coupled the corresponding negations, death, and darkness. Old and early as is light – not impossibly antedating most other cosmities or orderings of things – how new and late are not the conclusions of science respecting its physical basis! Light was only dissected yesterday, to discover all prior textbooks of its anatomy to be wrong. To-day no one questions the existence of luminiferous aether as a real substance, in the vibrations of which the quality of light subsists and is manifested. But this state of matter is impalpable, invisible, inaudible, inodorous, and insipid – in short, inappreciable to the physical senses. We know nothing about, as matter; we only know it is a mode of motion of matter in an unknown state. Force is obviously present and operative; matter is only an inference. But a substantial aether is a dictum of science, signed, sealed, and delivered. A vivid exercise of the imagination it must have originally been, and a lively act of faith in the evidence of things unseen, to set the matter before the reason, judgment, or critical faculty in such shape that the mind could not only affirm the verity, but be unable to deny the truth, as to the nature of light. How many men, in the history of intellectual achievement, are found capable of such splendid believing that they may understand – yet *credo ut intelligam* is required to unlock any of the great secrets of Nature, no less than is it necessary to penetrate the world of spirit. In the nature of the human mind such rational faith is the key to discovery. Imagination engenders, belief cherishes, observation nourishes, reflection educates, and judgment approves – then the result takes care of itself, as a mature scientific truth. The accepted theory of light, in simplest expression, is an unknown but believed-in state of matter in a known mode of motion – it is matter at an inestimable minimum of density moving with extraordinary velocity under a force of enormous intensity.

On the other hand, the grossly material basis of life is perceived by all experience – the body of any plant or animal shows what number and kind of known states of matter may be informed and instinct with the life-principle, among which are sand and lime and iron, and many others, besides those composing the supposed ultimate basis of life, protoplasm; while what amount of motion is imparted by what kind or degree of force has proven thus far inestimable. What may actually be the facts in the case, however, so far from being inconceivable, is to my mind a very thinkable proposition, with the possible truth of which no known phenomena of life are necessarily irreconcilable.

Thus, since I cannot imagine force primarily act upon matter in bulk – like kicking a stone – it is necessary to infer, for the validity of the vitalistic theory of life, an excessively tenuous state of matter set in motion by an excessively active force – just as I did when speculating upon the connection I imagined to exist between mind and matter. Such conditioning of matter and force would be strictly comparable to what is known of the nature of light. It would be the analogue of – perhaps the homologue of – possibly identical with – that inter stellar fluid which is recognized by science as the universal medium of transmitting energy. It would, however, differ from light in several important and essential particulars. To satisfy the conditions of the theory, the substance or physical basis of biogen would be perhaps as much more tenuous than luminiferous aether as is the latter more fluid than hydrogen; it would be at the actual minimum of density at which it is possible for force of any kind to be transmitted, and so operative and manifest. The velocity of motion would be only less than infinitely greater than the known velocity of light, for it would be at the rate of speed at which thought can be transmitted. And as to the kind of force which would effect such motion of such matter, it would differ from any kind generally recognized, in that it would be self-conscious; that is to say, it would be pure spirit.

According to the terms of my speculation, the vital principle is a real entity – an *ens realissimum*, the incorporation of which is protoplasm or any other combination of gross matter makes such matter "alive", and the dissolution of which from such matter leaves the latter "dead". Biogen itself, of course, is alive; it is life; and biogen may be defined as spirit in combination with the minimum of matter necessary for its manifestation. Biogen is simply soul-stuff, as contradistinguished from ordinary matter; [8] it is the

---

8

"Mind-stuff," and the "hypothetical substance of mind," are expressions already in current usage among scientific writes of

substance which composes that thing which a well-known and very frequently quoted write calls the "spiritual body".

I have spoken to little purpose, and my expressions have been ill-chosen, if what I have said seems novel to you; if you do not discover in what I have said simply a restatement, in somewhat "scientific" language, of one of the oldest, and I think one of the wisest, of the world's conceptions of the life-principle, as a direct effluence of the Deity. It is the old *anima mundi*, the soul of the world, "workshop of nature," where the will of God is first fashioned in form and substance to receive the breath of life. And it is instructive to note, that in the whole history of human notions respecting the origin and nature of life, the theory of spontaneous generation, which the strongest science of today most strongly disclaims, is the one which has taken hold upon the human mind. Biogenic speculation has almost invariably flowed in the stream which bears the idea of father and son upon its bosom. Let us not deceive ourselves with the giving new names to old things. Call them what you please – modern materialistic and atheistic notions about life are every one of them disguises of the plain statement that a self-created atom of matter lays an egg that will hatch. Call this a monstrous absurdity, an instigation of the devil, if you choose; I can call it neither science, nor philosophy, nor religion, nor anything that is learned, wise, or true.

---

repute. What is meant by these terms I cannot imagine, unless indeed it is that very real thing which I call "soul-stuff?"

Biogen

To my mind the *anima mundi* belief, as I restate it in terms of the biogen theory, acquires color from the consideration that it is exactly the complement, and perhaps the natural antinomy, of the generally received views respecting the evolution of chemical elements and chemical compounds from indifferent states of nebulous matter; and not unlikely to be quite as true. The progressive consolidation of matter, during which the most diffuse, most tenuous and indifferent substances as gradually differentiated and then combined to form the various products known as "elements", to be recombined in endless diversity to form "inorganic" and "organic compounds" – such process would seem to involve as its necessary conditioning of universal antinomy, that at a certain stage of molecular aggregation reached by certain forms of matter, the counteractive vital principle comes into operation to arrest the consolidation, to bring matter out of the depths of gross materiality it has reached to the sublimity of effectual contact with spirit. Whence emanated matter in the beginning is inscrutable; from nowhere, certainly- if not from the self-conscious, self-determining universal Mind which willed to so become manifest. Where to? Nowhere, certainly- if not to whence it came, to complete the circle, symbol of infinity, whose quadrature is unknown.

Equally unknown are the time, the place, the circumstances of the origination of life. We may learn of these things when we discover what is matter divorced from force; for of neither of these things, apart from the other, if they be not one in essence and that essence pure spirit, do we know anything at all. The vital principle, which I must certainly invoke to satisfy the fundamental data of my consciousness, is equally inscrutable; but it is peculiar, in that it is not known to be manifested except in consequence of itself, or to reside long in any one glomeration of gross matter, or to ever die. I am bound to consider it as the most direct and immediate natural manifestation we have of the Great First Cause, and consequently to refer it at once far back of any such secondary cause as a chemical or mechanical law. I cannot suppose it will ever be determined either to originate in protoplasm or any other material compound, or to permanently reside in anything that retains the least vestige of materiality. Being absolutely beyond the scrutiny of the physical senses, it would scarcely appear to fall within the scope either of science or philosophy; and I doubt that human reason , unenlightened by revelation, can learn much about it; for that would be to find out God by taking thought.

Since the retiring President of this Society has declared that neither science nor philosophy affords any foundation of proof upon which my conscious mind may build hopes of that immortality of the soul which is that same mind a necessary conditioning of its existence; it is to be hoped that science may yet discover facts enough, and philosophy find truth enough, to render that result possible; for until they do, they are together obviously incompetent to deal with the life-problem; and until they do, fellow-men must be permitted to interpret the great secret each after his own methods, as best suits his own necessities; even should these force him to take refuge in some credible formulation of faith, as in something which certainly promises more than science and philosophy have accomplished, and may contain the germs of a good working scientific hypothesis.

But there is science and science, more or less intelligent or intelligible. There is philosophy and philosophy – that of Socrates, and that of Kant, for example. In such wealthy embarrassment, the real lover of wisdom may be inclined to seek truth in ways that vex his mind least, and at least leave him at peace with is soul, ignorant though he be of its origin, nature, and destiny.

Here, gentlemen, I should cease speaking. But my speculations have been surrendered to your criticism; and, as I know that many colors are reflected in the mental spectrum of the philosophers present, I beg you, in the discussion about to ensue, to resolve my doubts in the following particulars:

What is the difference between a Godless, self-created, always-existent cosmos of matter-in-motion alone, and any perpetual motion machine which men have dreamed of inventing, but which philosophy declares impossible?

What is the difference between any mechanical or chemical theory of the origin of life, and that spontaneous generation of life which science declares to be known?

What is the chemico-physical difference between a live amoeba and a dead one? And if there be no chemical or physical difference, in what does the great difference subsist?

What is the principle difference between a living human being and his dead body, if it be not the presence or absence of the soul? And if it be nothing like this, what, then, is it more like?

Appendix

A man's "mind" is not a *thing,* in the ordinary sense of the word thing. It is a relation between two things. These two things are, his soul and his body. The mind is the result of the interaction between spirit and matter. It is what the spirit thinks in consequence of its connection with matter. It is the knowledge which the spirit acquires by its experience in contact with matter. It is what the spirit must become incarnated to discover and appropriate. It is what the spirit retains when it becomes disembodied. It is the knowledge of good and evil. It is the fruit of the tree of life.

Reason is the mistress of the mind, and its exercise is judgment or the critical faculty. But its data is only those which it receives through the avenue of the senses. The bodily senses are obviously and notoriously fallible. Reasoning upon such data as the bodily senses may give may therefore be equally deceptive; and thus the results of reason are often fallacious, though its processes may be perfectly logical. Hence what any man *thinks, i.e.,* his mind, may be very wrong indeed, since it is necessarily based upon experiences of his spirit with matter.

On the other hand, a man's soul is a thing, in a proper sense of that word. It is a substantial reality, an actual entity, a living being of knowable and recognizable qualities, attributes and potencies. It is not merely a thought, or an idea, or any metaphysicality. It consists of a kind of semi-material substance, which is the body of the spirit, bearing much the same relation to pure spirit that the physical body bears on the soul itself. The substance of the soul is the means and medium of connection or communication between spirit and matter. Soul-stuff is animalized astral fluid; that is to say, some quantity of the universal aether, modified by vital force, individualized by a man's spirit, and appropriated to the uses of an individual spirit, just as a certain quantity of grosser matter is individualized and appropriated to the formation of the physical body. The substance of the soul, to which I apply the name 'biogen', seems to correspond closely to what Prof. Crookes calls the 'fourth state of matter'; and some demonstrable activities of matter in this radiant state appear to be summed by him in the term 'psychic force'. It is the 'od' of Prof. Reichenbach, and many of the manifestations of its activities are grouped under the expression 'odic force'. It is what some appear to mean by the term "hypothetical substance of mind". It serves as an "aetherophore" – to borrow a word coined by Prof. Cope. One of its modes of motion was demonstrated by Galvani. The commonest and best-known exhibitions of its active agency are those of our bodily sensations and movements, its currents to and fro between a human spirit and that spirit's carnal envelope but described by described by modern psychologists as sensory and motor nerve impulses.

Some modifications of soul-stuff exist in all animals and plants – in all things which have life, if not also in those other things which we call inanimate. In the higher animals – in man at any rate – it becomes the vehicle, the envelope, and the instrument of spirit, indwelling in the physical body so long as his body is "alive", and leaving it is what is called dead, which is when the spirit entirely withdraws from the physical body, carrying its soul-stuff along. Thus a man, in this world and in the flesh, consists of three different and separate things. First, His physical body, certain transient atomic and molecular aggregations of solid, fluid, and gaseous matter. Second, His soul, a certain substance temporarily in contact and very intimate connection with his body, on the other hand, and with his spirit, on the other, serving as a medium between the two. Third, His spirit, of which he knows nothing, though his spirit knows itself perfectly well. "Death" is simply the disengagement of the third and second of these from the first. The deserted physical body, no longer animated by the spirit acting through the soul, is "dead"; it has lost its "vitality"; the "vital principle", which is simply the force by which the spirit acts upon matter through the medium of the soul, is no longer operative; and the body in this state, *i.e.*, dead, is only acted upon by physical and chemical forces. It then furnishes a very proper subject for the chemico-physical theory to explain and account for.

"Mind", as the expression of a relation between the soul and the body, necessarily disappears when that relation is discontinued. But a far higher order of intelligence, volition, and will-power is manifested by the spirit as soon as it is separated from the physical body. Having then a dual being only, instead of a triple mode of existence; replacing mere mental reason with those higher spiritual facilities whose glimmerings and fain foreshadowings in this life it used to call imagination" ; contrasting more clearly than it could while in the flesh the meanness of the intellectual with the majesty of the moral facilities; appreciating the great gulf fixed between good and evil; limited in its activities neither by the three dimension of space to which it was confined while in the body, nor by the modes of motion then known; - the human being has entered upon another sphere of existence by an evolutionary process as natural as that by which he passed from the womb to the world. The transition is probably less abrupt, in most cases, and there is no reason to suppose that the change is any greater. The body does not appear to be any more to the existence of the soul in the other world than is the after-birth to the existence of the body in this one.

From what has preceded it is evident that what I mean by "soul" is not exactly according to the general usage of the word; which usage commonly makes "soul" and "spirit" one and the same. Thus, when we speak familiarly of "a man's soul", we also say it is "his immortal spirit," meaning thereby, anything and all there is to a man which is capable of surviving death. But, as already stated, I draw a wide distinction between "soul" and "spirit". Spirit is nothing if not immaterial, and to "spirit" proper we can attach no significance if we do not consider it as divested of every trace of materiality. Soul, on the contrary, is substantial, and semi-material; it is the "body of the spirit", necessary, so far as we know, to all and every manifestation of the spirit. Spirit cannot act directly upon matter, but only through the intermediation of this soul-substance. A human being, after "death", consists of this substance, acted upon by his spirit, the two together constituting what is ordinarily called his "soul". To this substance, when acted upon by, and serving for the manifestation of, spirit, I give the name biogen. The same substance (biogen) acted upon by the spirit before death of the body, and serving for the operation of spirit upon matter, is the "vital principle", the action of which we call "vital force", and the results of which action we call "vitality" or "life".

I do not admit for the instant that biogen is merely an idea, or thought, of mine or any one else- a metaphysical abstraction, a mere mode of expression, or a mere mode of motion either. It is not, furthermore, a relation subsisting between two things. Nor is it a "force", in the ordinary sense of the term. It is a THING, a very real *thing,* an *ens realissimum,* possessed of sensible qualities and attributes which may be investigated by proper scientific methods, and by scientific experimentation, quite as readily as any other of the so-called "imponderables" of nature. It is as open to examination as luminiferous aether, and its properties, if not its substance, may be studied as we would study light, heat, or electricity; it is therefore not only a proper object of science, but a proper subject of philosophy.

Under ordinary circumstances, biogen is inappreciable to the physical senses, however manifest its effects. Under exceptional circumstances it acquires very sensible properties, the principles of which are visibility and tangibility. It may then be both seen and felt. Its modes of motion appear to differ in some respects from any of those known to us to be possible to gross matter, and to require for their complete exhibition more than three dimensions of space. Its excessive tenuity, extraordinary elasticity, compressibility, homogeneity and some other qualities, lead me to suppose that one great difference between biogen and most known states of matter maybe, that it is not of atomic constitution. If hydrogen, the most subtle and tenuous gas known, cannot exist in a free state except two of its atoms be joined in a molecule – and this is good found chemistry of the day – it may be that biogen consists of free atoms; that is to say, differs chiefly from other kinds of matter in having no molecular constitution. More probably, however – viewing some of its properties and activities – it is to be considered not even atomic in constitution- having no atoms of any size or shape or distance apart – no fixed points of greater density than their intervening spaces. In this view, biogen would be simply *tomic matter* as distinguished from atomic matter; and to so regard it may be well, for the present at least.

During the earthly life of the individual, a person's biogen appears to be normally confined to the limits of his physical body; or at any rate to make but faint and feeble excursions there from during his waking hours. In sleep, however, when the spirit is temporarily withdrawn from the outer world by the closure of the usual avenues of the senses, the biogen is much freer in its excursions, and may almost entirely leave the body at the will, consciously or unconsciously exerted, of the spirit. Probably no person "is himself" so much as in his dreams, under those conditions; a fact which Shakespeare doubtless knew, familiar as he was with the properties of biogen. When he wrote that we are such stuff as dreams are made of. More obvious through less familiar exhibitions of the excursions of biogen from its usual abode in the body are witnessed in various phenomena of somnambulism, spontaneous, or induced; in clairvoyance, clairaudience, trances of various kinds, religious ecstasy, some forms of catalepsy and epilepsy; and especially in what is called "suspended animation". Some persons are so constituted that they can project their biogen at will; others, that it flows from them unconsciously, against their will, during their waking hours; others again, that it can be drawn out of them neither by nor against their will, but under circumstances they have learned to recognize and to which they may voluntarily subject themselves. In highly exceptionally cases, frequently but not necessarily preceding death, biogen may proceed from a person in such quantity and of such quality as to be visible and even tangible to another person. At death, it's entirely withdrawn from the physical body, with more or less rapidity; and the act of dying is not accomplished until this process is completed, when the individual is at length dead, his spirits continues to live in a body composed of biogen; and this "spiritual body" may, and frequently does, become visible and tangible to those whose souls still inhabit their physical bodies. The substance which I call biogen, therefore, is an available, a legitimate and an appropriate object of scientific inquiry, by no means to be ignored in any system of philosophy, and by no means to be mistaken for protoplasm.

Part 2

Esoteric Osteopathy

Contemporary Preface to Esoteric Osteopathy

Remember the moments of first discovering yourself. Commonly, as teenagers, most of us felt the urge to reassess the nature of life and challenge the perimeter of human possibilities. So, too, each generation requires a renewed expression of the perennial truths of human existence. Yet as the transition to adulthood solidifies into identity, relationships and employment, the relevance of this quest, and the passion, the urgency and its extent fade for some, but adapt a more determined form for others.

And so, in each era there is a struggle over priorities between the majority who are content to operating uninterruptedly within the dominant social paradigm and those looking for a more complete or authentic expression of the truth. This latter group has spawned generations of insight to evolve and enrich human culture. Whether we call them Essenes, Gnostics, mystics, metaphysicians, alchemists, Theosophists, esoterics, beatniks, hippies, Aqaurians or hiphoppers, the dialectic goes on and human cultured is enriched.

Often, the reflective challenge is neither appreciated nor welcomed; it appears impractical to many.

A curious work

A.T. Still, the founder of osteopathic medicine, certainly had this spirit of challenging the status quo. The response of his students varied in this regard.

While browsing the back room of the library of the Kirksville College of Osteopathy some years ago I came across this small book, *Esoteric Osteopathy,* by Herbert Hoffman. The small work presents a reflective, apparently divergent view among Still's early followers. The date of publication, 1908, suggests a close tie to the thought of Still. But what can be esoteric about Still's biomechanical approach to health, disease, and medicine? Osteopathy is very practically oriented. This is the way osteopathy is taught, as a reasonably complete system of diagnosis and treatment focused on anatomy and biomechanics.

And in reading the text, the connection to Still is not immediately apparent, unless one takes seriously what appear to be non-biomechanical digressions in Still's writings. Here the reference is to his passages on biogenic life force and divine intention reflected in anatomy. But the relevance of Hoffman's writing, as will be further expand later, lies in the observation that Still's world view was always tempered by his quest for a larger vision of human existence and the reciprocal roles of mind and body in this contexts. The empiric, the intuitive, the spiritual - the unknown - shadow each other in Still's world.

Osteopathy has been claimed conclusively and univocally to support the philosophical and then alternately the practical approach to health and human existence. I believe Still's osteopathy claimed the complementary value of both approaches. However, in the current climate of "evidenced based medicine" or parity with mainstream medicine, the tendency in osteopathy has been to recognize the practical or defensibly scientific aspects of the tradition and downplay the more reflective aspects. In this context, Herbert Hoffman's *Esoteric Osteopathy*, and the importance of mental connectivity, appears to stands out as an orphan in the shelves of the archives. But, as Hoffman himself quotes Andrew Taylor Still, there is a resonance with the more philosophical streams of thought in the writings of the Old Doctor. Admittedly the connection is more in the spirit and substance of the writing of the two men and not their common terminology. But in my view, literature presenting this complementary view of life and osteopathy warrants review and integration into osteopathic education and practice. It helps recreate an appropriate balance of tensions between these positions.

Hoffman, in context

Although the fact and year of Hoffman's graduation has not been verified, the publication date of his work puts him in the era of rapid development and struggle for political credibility of osteopathy. In the early twentieth century, Still's energies were heavily taxed in defending his key concepts, managing a medical school and hospital, and politically consolidating a base of support. There was a lot of practical work to do and the need to convince a wide audience of the value and legitimacy of this new approach to medicine.

In analyzing Still's writing of this era, one recognizes that many of his key concepts in diagnosis and treatment revolve around the primacy of anatomy in medicine. The emphasis, especially when one reads other early works interpreting osteopathy (Barber 1898, Hazzard 1899, 1905), was on the anatomy as a material organization of parts and interactive systems. Much of the 100 year plus legacy from this early period reinforces this trend. Indeed, this is practical since it is on this level that one confronts challenges from the natural sciences, which medicine in general has adapted as its guiding beacon in assessing truth.

Complementing this anatomic and empiric theme through all of his writing, Still weaves another theme which may be used as a lens to see "anatomy" in a different way. He rather consistently refers to anatomy, and the body, as the work of an intelligent creator;

"Finding health", Still's goal in treatment, is a matter of reading the intent, the mind, of the Creator in the organized structure and function of the body.

"To find health should be the object of the doctor. Anyone can find disease."   Still 1899, p. 28)

"Man represents the mind and wisdom of God to the degree of his endowments." (Still, 1892, p. 27)

But treatment involves working consistently with "the intelligent God, who has formulated and combined life, mind, and matter in such a manner that it becomes the connecting link between a world of mind, and that element known as matter." (Still, 1899, p. 223)

This second theme reflects a different paradigm for osteopathy and is the key theme linking Hoffman's writing to Still's point of view. The living body is greater than the sum of its material parts. This totality includes aspects not typically engaged by unreflective palpation or visual observation. In this line of thought, Still adapts the concept of biogenic life force to help explain the vital, intangible aspect of the living person. Elsewhere he refers to the trinity of body, mind, and spirit to describe the scope of activity of this life force.

"First, there is the material body; second, the spiritual being; third, a being of mind which is far superior to all vital motions and material forms, whose duty is to wisely manage this great engine of life." (Still, 1892, p.16)

And while Still admonishes the osteopath to be a philosopher, "a seeker for the truth", even while he prompts his students to be engineers and mechanics; they are to appreciate a broader world view. The following sequence of excerpted quotes develops this theme.

"My object is to make the osteopath a philosopher, and place him on the rock of reason." (Still, 1992, p. 20)

"One of the greatest questions, if not the greatest, that has ever presented itself to any philosopher in any age is, what is life? ...At the end of all his philosophical labors the philosopher concludes that life is a substance and superior to the sum total of the elements of the whole universe. Its superiority is proven by one of its attributes which is mind. Mind by its unlimited skill rules, governs, and uses at will all forces and elements."

"Life is a substance which fills all the spaces of the whole universe."

"Life is the God, the wisdom, the power, and the motion of all."

(Still, 1992, p. 278, 279)

But few of Still's students appear to have heard this aspect, or to be able to find a practical direction in which to pursue and apply these further integrative concepts. This charge can challenge one's world view on a fundamental level, and requires analytic skills that many people do not want to cultivate. It requires the apparent expenditure of energy without tangible reward.

Herbert Hoffman accepts this challenge and his work represents the earliest effort known to this author to expand the scope of Still's expression of osteopathy in this direction. Later contributions by Charlotte Weaver, William Sutherland, Rollin Becker, Bob Fulford, Jim Jealous and Pierre Tricot continue to reflect that osteopathy's approach to function is broader than physiology as defined by conventional empiric biology. Often these approaches are challenged as not being congruent with the current consensus of scientific knowledge in which osteopathy is most often interpreted and therefore constitute quackery. (Weaver, 1938; Sutherland, 1967: Becker, 1997; Fulford, 1996; Jealous; Tricot, 2002

And yet, in our day, there is a swell of popular inquiry and interest in the area of complementary medicine, including many approaches dependent on alternative views of science and life, including the relevance of consciousness.

The current republication of *Esoteric Osteopathy* is intended to verify that the dialectic of understanding, between the tangible and the less tangible, the physical and the metaphysical, is perennial. It is part of the eternal struggle of human understanding to assimilate what is at the periphery of its comprehension, though conscious, and pull it into the circle of clear knowledge.

Deeper into Still's mind

As cited above, in his *Philosophy and Mechanical Principles of Osteopathy*, Still dedicates an entire chapter to his reflection on the essence of life, in terms of a biogenic life force. Favoring a vitalistic approach, Still apparently borrowed this term from the thought of Elliot Coues, an American ornithologist with a philosophical interest. (Stark, J, 2003, Coues, E, 1882) who using the term describe the nature of life in the context of general biology. In this referenced address before the Philosophical Society of Washington, Coues describes:

"Life in the concrete is, of course, the sum of the phenomena manifested by animated nature. Of life in the abstract, of the essence or nature of that peculiar attribute of plants and animal, apart from its material manifestations, no knowledge whatever seems possible. Yet, while I cannot even imagine what life is or may be, apart from matter, so far is it from being impossible to me to conceive of life as an existent reality apart from any *known condition* of matter, that it is impossible for me to not form that conception. This is of course to invoke the 'vital principle', to postulate the reality of a kind of force called 'vital', as a veritable Biogen or life giver, which may be where no known form of matter is, and can, therefore exist apart from such matter, and not as a resultant of any material force. Though this is pure speculation, I am forced to speculate, in the impossibility of conceiving the contrary." Coues, 1984, p 38

Coues challenges that his hypothesis is as credible as the proposition of the self-organization of matter from cosmic nebulae to single cell organism to the complexity of a human. Although he intentionally spares his audience the argument from religion, he admittedly relies on the classical chain of causation which relies on an ultimate or creative cause equivalent to the multicultural appreciation of God. He cites his coinage of the word "biogen" and he seems quite compatible, in all ways, to the thought of Still. The date adds to the credibility that his ideas influenced Still, just as did the thought of Herbert Spencer as we will see below.

Curiously, in another work, describing the biology of birds, more particularly the sense modalities of birds (vision, hearing, taste), Coues makes the following statement:

"All animals are probably also susceptible of *biogenation,* which is the affection resulting from the influence of biogen; a substance consisting of self-conscious force in combination with the minimum of matter required for its manifestation." (Coues, 1927, p. 198)

This statement if generalized can be very compatible with current concepts of bioenergetic palpation. (Comeaux, 2002)

Scope of osteopathic inquiry

This quest to define osteopathy is a participation in the human philosophical quest that has proceeded for thousands of years; Still's interpretation has been used to support the emphasis on both the tangible, controllable empiric aspects of life (which is categorized as the physical sciences), as well as the less tangible aspects of human experience (categorized as the esoteric or metaphysical sciences). Herein lies a tension between the known, and the unknown, the empiric and the more mysterious. In another speech usually not appreciated in this context, Still refers to his struggle to understand life issues and the nature of the person who is his patient. He discusses the apparent mystery of life and death:

"When I looked up the subject (life) and tried to acquaint myself with some of the works of God, or the 'Unknowable'...

Take the hand of man, the heart, the lung, or the whole combination, and it runs to the unknowable. I wanted to be one of the Knowables.

The first discovery I made was this: that every single stroke of God came to me as the unknowable. The stroke of death- what do you know about it? I know nothing, therefore it is unknowable." (Still 1981, p. 241)

In his discussion Still, the practical anatomist and inventor, expresses respect and urgent interest in the frontier between physics and metaphysics science and religion. Often not appreciated, Still is actually dialoging with an important influence on his thought, Herbert Spencer. (Trowbridge 1991, p. 117)

Spencer, a popular nineteenth century philosopher, was describing his position on the perennial argument between a materialistic and a spiritualistic world view. This was important business since he, Spencer, was a systematic philosopher, in other words, his philosophy dealt with the application of general principles to all aspects of life. And, so he has works on speculative philosophy, but also on physics, biology, sociology, and other topics, modeling all interaction according to the principles of interactions modeled on Newtonian physics. (Spencer, 1864)

In this context he actually describes the relationship between the Knowable and the Unknowable, as referenced by Still.

The pertinent point here is that Spencer begins his *First Principles*, an approach to scientific knowledge, by discriminating the Unknowable, "an absolute which transcends not only human knowledge but human conception", from the "Laws of the Knowable- A statement of the ultimate principles throughout all manifestations of the absolute." Still's language identifies that he is aware of this discussion, and attempts to reconcile the two through the concept of the body as part of the expression of the God of Nature, though made comprehensible.

But Spencer holds that neither spiritual enlightenment nor empiric study allows one to totally comprehend the nature of life.

"Perceiving as he will, that the materialist and the spiritualist controversy is a mere war of words, in which the disputants are equally absurd – each thinking he understands that which is impossible for any man to understand – he will perceive how utterly groundless is the fear referred to. Being fully convinced that whatever nomenclature is used, the ultimate mystery must remain the same, he will be as ready to formulate all phenomena in terms of matter, motion and force, as in any other terms, and will rather indeed anticipate that, only in a doctrine which recognizes the unknown cause as coextensive with all orders of phenomena, can there be a consistent religion or a consistent philosophy." (Spencer, 1864, pp. 481-2)

"He will see that though the relation of subject and object renders necessary to us these antithetical conceptions of spirit and matter, the one is no less than the other to be regarded as but a sign of the unknown reality which underlies both." (Spencer. 483)

Both Still and Spencer, as did Coues, grope for a way of relating to this "unknown reality that underlies both."

At the end of his chapter on biogen, or vital force, Still concludes with this statement:

"We have given a few thoughts on this line of life, hoping the osteopath will take up the subject and travel a few miles farther toward the fountain of this great source of knowledge and apply the results to the relief and comfort of the afflicted who come for counsel and advice." (Still, 1892, p. 258)

As will be seen in the reading of the text, Hoffman seems to have taken the Doctor at his word in this regard.

Hoffman's *Esoteric Osteopathy*, and its treatment of the importance relevance of mind (individual consciousness) and Mind (absolute consciousness of a Creator) is offered as a contribution to this perennial effort.

Hoffman's other inspiration

Surprisingly, little is known of Herbert Hoffman, the man, and how he came to osteopathy. Much more, however, is known of his inspiration, named in the dedication, Ramacharaka. Living from December 5, 1862 to November 22, 1932 this individual was a prolific writer publishing a number of books in the New Thought and Mental Science movements at the turn of that century. Beginning in 1902, he published 13 volumes of practical philosophy, psychology, and self improvement independently and through the Yogi Publication Society in Chicago. Yet the name Ramacharaka was an adopted pseudonym for William Walker Atkinson, born in Baltimore, accepted to the bar in Pennsylvania and Illinois, and functioning as an attorney and business man. It is hypothesized that Atkinson adopted the name either to shed respect on a previously revered historical figure, or to separate his literary identity from his business.

It seems likely that Atkinson, in the course of his work in New Thought, became interested in Hinduism and received instruction from Baba Bharata, a pupil of the aging Ramacharaka (1799-~1893). The particular facts are difficult to substantiate. Much more detail and discussion is incorporated in the web references below.[9]

So what were Ramacharka's key ideas and how are they relevant here? How does Hoffman use them? How do they relate to the ideas of Dr. Still? How are they relevant to us?

Atkinson presents a recapitulation of Vedantic or Hindu panpsychism. In other words, the underlying substrate to all reality is mind or Mind. In this system, the absolute Mind is creative of all that is. Individual mental activity, in healing or in general social life, is causative. Therefore, thought and its expression in various forms are interactive, not only within its own domain, but also on the physical plane. Both mental and physical reality are expressive of the same force. Therefore, thought can and routinely does create change in the physical world.

"all that we call Matter (or Substance) and Mind (as we know it) are but aspects of something infinitely

---

[9] http://williamwalkeratkinson.wwwhubs.com

Ananda_ji@gmx.net

higher, of which may be called the 'Cosmic Mind' ".
(Atchinson, 1997, p. 27)

"...we believe that the secret of Health lies in the observance of the Nature Laws of the Body. The Laws may be summed up as Right Living and Right Thinking."

(Atchinson, 1909, p. 16)

"Vital Force is something which plays its own part in the economy of Nature."

"All persons have more or less Vital Force, and all people have the power of increasing their store, and of transmitting it to others, and thereby curing disease."
(Atkinson, 1909, p. 49)

In a Vedantic approach, the individual mind(s), in its root nature, is participative and an expression of Absolute Mind. Therefore, a variety of disciplines for breathing (pranayama) , physical training (Hatha yoga) and meditation are approaches to create harmony of the individual body and mind, as well as conforming the individual to the Absolute. Ramacharaka recognizes and describes the role of the mind in physical, mental and spiritual healing.

This seems to mirror heavily the intent of Still to recognize the expression of the Creator's intelligent plan in the organization of anatomy and physiology. Still's quest for health and an approach to healing seem to parallel these concepts. However, a primary difference in Still's approach is that he encourages his followers to accept the challenge, and the responsibility, as a healer, to facilitate the reconciliation of the patient in pain or disease with the

Creator's intent. Part of the exigency comes, no doubt, from Still's initial motivation to prevent recurrence for others of the personal loss such as he had experienced with the loss of his wife and children to meningitis. So Still developed strategies and principles to be applied in the context of disease. Immediately, this meant assisting the body in its return to ideal or unimpaired function. This included function and free mobility of the musculoskeletal system, a harmony intended by the Creator, but Still included unimpaired vascular, neural and other vascular activity as part of function. What would have constituted the fullness of Still's thought in his later life, in reconciling the diverse components of his philosophy, is a matter of speculation.

Hoffman's approach

Hoffman, then, sees Ramacharaka's synthesis as a bridge between the mental and physical aspects of Still's thought, and a way to reconcile with Spencer's concern about the absolutely unknowable. He addresses his attention and strategies to leading the osteopath to make similar connections, in redefining the patient, their illness, and the scope and strategy of engaging the patient to heal.

And so, as Still, Hoffman directs us to change our attitude, our intention regarding treatment and to engage the patient in a different way. Specific technique is not the emphasis. Hoffman most importantly describes the intention to engage not just the form and organic function of the body and its community of organs, but also to engage the mind of the individual parts.

"If you have a hard, contracted muscle, just work *easily* over it, telling the 'Muscle Mind' to relax, and it will do it in a very short time, thus doing sway with the 'back-breaking,' strenuous work that you have been accustomed to doing."

"Now in closing we will ask you to always bear in mind, that no matter what disease you may be called upon to treat, to remember that it is produced by *imperfect* 'mind action.' Give your usual Osteopathic treatment, plus the *mental commands* in words suited to the case, and you will be practicing Esoteric Osteopathy in Reality. That is all there is to it."

"Above everything else, remember you are talking to the MIND of the organ or part, not to matter. Also always remember that there is no *dead matter* in a live body. Mind is in every part and cell, dead though its matter seems." (Hoffman, 1908, p.27)

This clearly is an extension in the osteopathic context of Atkinson's work.

"The principle of mental healing lies in the fact that the central mind controls the bodily functions- or the mind manifesting through the organs, cells, and parts of the body." (Atkinson, 1909, p. 157)

Following the thread

Hoffman was not alone in looking beyond basic biomechanics in osteopathy. Although very early William Sutherland's interest began with subtle biomechanics, through his life he reached into other dimensions, including reaching for an understanding of a deeper Potency driving life in the individual. Several of Sutherland's students, among them Rollin Becker and Robert Fulford took these concepts in different direction. Becker had a unique way of minimizing theory and maximizing the experiential approach to interacting with the dynamic state of the patient.

"A note in listening: ...In the process, stop *thinking* about it and surrender to the total anatomicophysiological output of the patient." (Becker, p. 149)

"At the very core of total health is a potency within the human body manifesting itself in health. At the very core of every traumatic or disease condition is potency manifesting its interrelationship with the total body in trauma and disease." (Becker, p. 165)

"By following the biodynamic intrinsic forces and their potency and the biokinetic intrinsic forces and their potencies through the potency or stillpoint within the tissue pattern within the patient, I have been able to secure therapeutic benefit for the majority of pathological conditions encountered within the patient." (Becker, p. 176)

This means of working remains difficult to describe. Anthony Chila, a student of Becker, describes this phenomenological approach to palpation and treatment. (Gaines, Chila, 1998) Becker is also an inspiration for James Jealous, who incorporates his mentor's term for the body's inherent vital behavior, or biodynamics.

Esoteric Osteopathy

Another of Sutherland's students, Robert Fulford, followed the thread of vital force and expressed a correlation of Still's trilogy of body, mind, and spirit in three major ways. Philosophically he explored a cosmology in which primordial Creative Energy underwent a stepping down in intensity from the Creator, through love expressed physically as light, through thought expressed in mind, progressively into the subtle and more material aspects of individual personalities with material bodies. Anthropologically, these and similar concepts are expressed in various cultures. Fulford was inspired by Randolph Stone and Edwin Dingle who incorporate the oriental view of life including an emanation theory similar to that described in the book here reviewed. This emanation, if reversed, describes the life quest for reunification with the Absolute which is the purpose of life. Although this quest may persist through multiple lifetimes, health reflects an unimpeded developmental process; disease or pain reflects impeded progress. In this tradition, the physical body was sustained by the less tangible etheric body, which in turn related to the mental and spiritual body. (Comeaux, 2002)

In Fulford's synthesis, these anthropological and philosophical approaches are in the process of being validated by scientific experimentation. This represents the third theme in his thought. Fulford felt that there existed a direct current field which was the physical manifestation of the etheric field and was an intermediate, organizing principle for the body. The key osteopathic feature to Fulford's approach was that these subtle relationships were, with concentration and practice, palpable. Furthermore, through guided intention, the energy of the practitioner is transmuted to interact with these relationships, which constitutes a form of manipulation.

This quest to more fully understand the nature of the patient and the means and limits of our interactions goes on.

Summary

Still overall had a respect for the physiological and structural order of the body, and their interrelationship. Additionally he wrestled with the inclusion of spiritual and vital nature of the person. And yet the expression of these aspects and their interrelationship came difficultly, and was more cumbersome to integrate cleanly into the more acceptable biomechanical model.

Hoffman, as a contemporary of Still's, presents an attempt at applying another model, bridging the gaps between Still's "material body, spiritual body, and the body of mind".

In bringing this topic to light, in rereading and republishing this old manuscript, my intention is only to validate the awareness of those who have already made the connection between palpation and deeper insight, but also to tease those who already "understand it all" without making this connection to loosen the perimeter of their worldview and to learn more. In either case, the result is discussion and dialogue, and possibly an expansion of the frontier in applying osteopathic thought in practice.

References:

Trowbridge, C., Andrew Taylor Still 1828-1917. The Thomas Jefferson University Press, Kirksville, MO, 1991, p. 117

Spencer, H., First Principles, 5th edition A.L.Burt Publishers, New York, N.Y, 1880 orig. 1864

Hazzard, C., Principles of Osteopathy, Detroit, by author, 1899

Hazzard, C. Practical and applied therapeutic of osteopathy, 1905

Barber, E. Osteopathy Complete, Kirksville, MO, Hudson-Kimberly Publishing Co.1898

Still, A.T., The Philosophy and Mechanical Principles of Osteopathy, Osteopathic Enterprises, Kirksville, MO, 1986, orig. by author, 1892

Still, A. T. Philosophy of Osteopathy, Kirksville MO, 1899

Still, A. Research and Practice, Eastland Press, Seattle, 1992, orig. by author, Kirksville, MO 1910

Still, A.T., Autobiography of A.T. Still, American Academy of Osteopathy, Colorado Springs, 1981, orig. 1908

Stark, J. E., Still's Fascia: A Qualitative Investigation to Enrich the Meaning Behind Andrew Taylor Still's Concepts of Fascia. Jolandos Pub, 2003

Coues, E., Key to North American Birds. 6th edition, Boston, The Page Company, 1927, orig. 1872

Coues, E., Biogen, a speculation on the origin and nature of life, 2nd ed, Boston, Estes and Lauriat, 1884

Atkinson, W., The Science of Psychic Healing, Yogi Publication Society, Chicago, IL 1909

Atkinson, W., Dynamic Thought or the law of Vibrant Energy, orig. 1906, republished by Kessinger Publishing, Whitefish, MT 1997

Gaines E., Chila, A., 1998 Communication for osteopathic manipulative treatment (OMT): the language of lived experience in OMT pedagogy, Journal of the American Osteopathic Association vol 98, 3

Becker, R., R Brooks ed., Life in Motion, Portland OR, Rudra Press, 1997

Fulford, R., Dr. Fulford's Touch of Life, Simon and Schuster, Inc N. Y. 1996

Comeaux Z., Robert Fulford DO and the Philosopher Physician. , Seattle WA, Eastland Press 2002

Tricot P., Approche tissulaire de l'osteopathie – un modele du corps conscient Sully Vannes cedex France, 2002

Jealous, J., Emergence of originality: a biodynamic view of Osteopathy in the Cranial Field, self published Franconia New Hampshire.

Sutherland, W.G., Contributions of Thought, The Sutherland Cranial Teaching Foundation, 1967

Weaver C., The Three Primary Brain Vesicles and the Three cranial Vertebrae, 1938 Journal of American Osteopathic Association 37(8) pp.345-350

ESOTERIC OSTEOPATHY

—BY—

DR. HERBERT HOFFMAN

Philadelphia
DR. HERBERT HOFFMAN

1118-1120 Chestnut Street

This little book is lovingly dedicated to Yogi Ramacharaka, that illuminated soul, who led me from darkness to light.

The Author.

Greeting

"Truth is within ourselves; it takes no rise
  From outward things. whate'er you may believe;
  There is an inmost centre in us all,
  Where truth abides in fullness; and around
  Wall upon wall, the gross flesh hems it in,
  This perfect, clear perception—which is truth."
  Browning.

This little book is sent forth with an earnest desire to spread the light of truth, and to enable Osteopathic physicians to more effectually and understandingly cure the ills of their fellow men and women.

The truth as set forth in this little manual is not original with us, but has been handed down from the Eternal Source through various channels, and we just helped ourselves to the good things whenever and wherever we found them. Now it is our earnest hope that you who are reading these lines will also help yourself to whatever appeals to you. We have no desire to thrust upon you anything that does not appeal to your reason. Therefore, if anything set forth in this work does not suit your mental tastes, just pass it by, and take that which does.

We shall not attempt to prove any of the statements of truth made herein, but will simply recite what they have proved to us in our work as a physician. We possess no power that you do not also possess. You are a sleeping God, and need but the help of your own earnest thoughts backed by your own will to awaken you to a realization of your powers. Of course, if you have awakened to a knowledge of metaphysical laws, you will more readily grasp and use the healing methods set forth here-in, but even this is not absolutely necessary, if you but have faith and apply the methods which we shall give you. "The proof of the pudding is the eating of it," therefore as you progress in the application of these methods, you will be astonished (at *first*) how readily you will get results.

We would advise all who wish to further understand the truth, about the real laws of their being, and their exact place in the universe, together with their relations to the source of *All Power,* to commence by reading Yogi Ramacharaka's *Fourteen Lesson, in Oriental Occultism*, published by the Yogi Publication Society, Masonic Temple, Chicago, Ill. This book can be obtained for the very reasonable sum of one dollar. The book states the most profound truths, and hidden mysteries, in plain, simple English. No one can read it without being filled with a deep love and an abiding sense of reality. No matter what your religious beliefs, this philosophy will not run contrary to it, but will go along side by side, and enlarge upon it. It teaches of the true brotherhood of man, and the end and aim of all—union with God.

One who unfolds into a true knowledge of the "I Am" consciousness, and his connection to the "Great I" has at his command spiritual powers of healing, which heal the worst pathological conditions almost instantaneously. Therefore, if you would follow in the footsteps of the Son of Mary, begin by studying the true laws of your being as set forth in this pure and forceful book. After carefully reading and digesting the contents of these fourteen lessons, and you find that they supply a longfelt want, you may take up the higher courses, given by the same author, in the order named, "Advanced Course," "Raja Yoga," "Gnani Yoga," "Hatha Yoga" and his book, "Psychic Healing."

We are giving you here at one stroke the means to obtain knowledge, that took us many weary months of hard searching to uncover, not to speak of the useless time and money spent in wading through piles and piles of chaff and theories as advanced by many writers on the subject. But we do not regret the time spent, for it has taught us that true knowledge *comes from within,* and that the best that any writer or teacher can do for us is to awaken thoughts that are lying loose and disconnected in our own mind. They but confirm our own thoughts. Give us the loose end of the ball, and we unwind at our leisure.

We have shown you the way. It is now for you to decide which path to take. Here is the way— choose. One path leads to true healing knowledge and power, and is based on Reality. The other is based on shadowy and seemingly real things—on surfaces and outward things. You see the shape, size and condition of the surface, and think *it* the real thing, the final analysis. The underlying Reality is hidden, is not evident to the senses, cannot be weighed, measured, tested and demonstrated on a backboard. Then why should you bother with something you can't see. Ah, that's it. Why should you? Well, decide for yourself.

We will now bring this little greeting to a close, and go on and give you our method of "Esoteric Osteopathy". Try it. Use it. If you succeed in dispelling some nasty pathologic condition that has resisted your best efforts, and are thereby enabled to bring health and happiness to some poor sufferer, then we shall feel well rewarded for putting forth this humble effort to spread the glorious light of truth.

DISEASE—WHAT IS IT?

Before we attempt to cure, we must understand what we are attempting. If we find an organ or part, functioning abnormally, we call it disease, because it is not functioning as Divine mind originally intended it should.

Our text-books give many causes, outside of the organ or part, as the beginning cause of the disease. Not one teaches *us* that disease has its beginning in a disturbance of the "MIND" of the organ or parts. Not one even tells us that the organ or part has any such a thing as "Mind." True, histologists tell us that the cell has intelligence, but they stop there. Histologists never once dream of telling us that that intelligence is *"mind."* But reason tells us that there can be no intelligence without mind.

All physiologists inform us of the wondrous work of our bodies in health and sickness. Tell us of the constant work of repair, replacement, change, digestion, assimilation, elimination, etc., etc. And, most wonderful of all, the selective action of the cells extracting from the blood the nourishment needed, and rejecting that which is not needed. All this they tell us. And, more marvelous still, they tell us of the healing of wounds, the rush of the cells to the point where they are most needed, and hundreds of other examples well known to students of physiology. But of the most wonderful and marvelous fact of all they tell us nothing. Back of all this "MIND," the *mind* from which every cell expresses intelligence. Now, then, a cell has "MIND." It goes on reproducing itself many million times until there are many millions of cells that have amassed themselves, or built themselves into an organ, say into a Liver, by the way of illustration.

Now, then, this group of cells co-operating from the beginning to build themselves into a liver, must have

105

been of *one mind.* If this was not so, then some cells of that group would have started to build themselves into a heart or a nerve, or a blood vessel. But they didn't; they combined as one mind, having but one purpose, to make themselves into a Liver. So, then, we have a group mind called the *"Liver Mind,"* which mind thoroughly understands how to operate or carry out its own function, of course always under the supervision, direction and control of the central mind, located in the brain. To further clinch the matter, how could the central mind give orders to the Liver, if it did not have a mind to receive and execute the order?

If, then we have a "Liver Mind," Then we have also a "Heart Mind," a "Stomach Mind," a "Kidney Mind", a "Lung Mind" a "Solar Plexus Mind," and so on throughout the body until every organ, part, ganglia, plexus, nerve and blood vessel has this. *"Mind."*

Then all pathologica1 conditions have as their first cause, *"MIND."* Without "Mind" there could be no action in matter.

Then ALL disease is MENTAL.

Even diseases produced by violation of physical laws have their first cause in mental action. It is a mental action that *first recognizes the violations.* If a stomach *is* overworked, what tells it that it is being imposed upon? Ans.: "Mind." The "Stomach Mind" reports to the "Central Mind" that an oversupp1y of supplies is being sent constantly to it. The "Central Mind," realizing that if this is not stopped the entire physical body will soon be overloaded, orders the "Stomach Mind" to cease receiving any further supplies, and the Stomach obeys. Then we have resultant disease of the stomach, all of which are **familiar** to you. Even the act of the violation itself had its beginning in a *mental desire.*

Look at it from any point of view that you will; go around and around, and you will finally have to admit, that if the matter of which our physical bodies is composed, is changed, or shaped normally or abnormally, *there must be mind intelligence back of the action.*

Disease, then, is produced by wrong mental action. Therefore disease is abnormal, being only the result of temporary departure from right mental action.

It is not an entity. It has no principle and no intelligence of its own. Neither does it possess power. Man in his essence is whole and it is no part of him.

The word "nothing" contains, the root meaning of the word disease. It was derived from the Latin *dis* meaning un: without: the lack of; and the word ease. It stands for absence, not presence. It cannot exist without consciousness. The unconsciousness cannot be said to be even uneasy.

So this disease must be a matter of "Mind" only and a thought-condition.

It depends entirely upon either the conscious or the sub-conscious action of mind for existence, even in appearance.

Set the mind action right and you will re-establish the law as laid out in the beginning, when all things were manifested, and established whole and good by their Manifestor.

Health is the normal condition of humanity, but it must be maintained in the mind-realm, or it cannot prevail in the bodily representation of man. Let us think health and have full confidence in its *Reality* with as much spontaneity as in the past, we have appeared to think disease, through our fear of it, and see if we do not find mind a true healing influence, and a soothing lotion for the many seeming ills of daily life.

Let us begin to-day to exercise our minds to see only health; let us begin at once to refuse to accept the words of sickness and trouble, for they but express wrong mind action in matter. *Mind is all.* Everything that you see, that looks tangible and solid, is but an expression of mind. A lump of clay is an inert shapeless mass until the hands directed by the mind of the artist shapes it into a beautiful statue. The musical instrument is silent and cold until the mind and hands of the composer play upon it. A pile of stones is built into the beautiful edifice, first conceived in the mind of the architect. The great oak came from the mind in- the acorn. The Brain and all the rest of the tissues of the body resolve themselves back to the universal atoms of matter when mind departs at the "death" of the physical body. Do not make the deplorable mistake of thinking that the Brain secretes mind, like the Liver secretes bile. The brain is just as inert as the clay or stone until mind works through it.

The MIND and YOU, dear reader, existed before the physical brain and body, and will continue to exist long after its atoms have turned to dust. And remember this, for it is a *great truth.* You, the *Real You,* are *absolute master of mind, energy and matter.*

Hear what Victor Hugo has to say about man's powers. How beautifully he pictures *your greatness* over the body that you are now living in. He tells you that *you are a soul,* having a body, not a body having a something called a soul. He says:

"Man is an infinite little copy of God; this is glory enough for man. I am a man, an invisible atom, a drop in the ocean, a grain of sand on the shore. Little as I am, I feel the God in me, because 1 can also bring forth from out of my chaos. I make books which are creations; I feel in myself the future life; I am like a forest which has more than once been cut down—the new shoots are stronger and livelier then ever. I am rising, I know, toward the sky. The sunshine is on my head. The earth gives me its generous sap, but heaven lights me with the reflection of unknown worlds. You say the soul is nothing but the result of bodily powers. *Why, then, is* my *soul more luminous when my bodily powers begin to fail.* Winter is on my head and eternal spring is in my heart. There I breathe at this hour the fragrance of the lilacs, the violets and roses as at twenty years ago. The nearer I approach the end the plainer I hear around me the immortal symphonies of the worlds which invite me. It is marvelous yet simple. It is a fairy tale, and it is history. For half a century I have been writing my thoughts, in prose and verse, history, philosophy, drama, romance, tradition, satire, ode and song. I have tried all, but I feel that I have not said a thousandth part of what is in me. When I go down to the grave I can say, like many others, I have finished my day's work; but I cannot say I have finished my life. *My days will begin again*, the *next morning.* The tomb is not a blind alley; it is a thoroughfare. It closes on the twilight to open on the dawn."

The central theory of Esoteric Osteopathy is that dis-ease is a MENTAL trouble—not a trouble in the central mind, but in the *mind in the parts.*

The alert reader will now pause and ask the question, "What originally caused the trouble in the *mind* of the parts**?"** Now that is a mighty good question, and on our ability to answer it, depends the causation of all dis-ease.

First, we will take violation of physical laws as one cause. These laws are all well known to you; therefore we will not take up these pages repeating them. We might state *in* passing that you can very readily see how an overworked stomach could cause the "Stomach Mind" to rebel and go on a strike for a vacation. The same could apply to the "Kidney Mind." Or again, if the "Stomach Mind" could not obtain sufficient and proper food from which to intelligently make blood containing the proper nutriment, that its own cell bodies, as well as the cell bodies of every other part of the physical body, would suffer from starvation. If mind is to manifest normally in the cell it insist have a body that is perfectly nourished, otherwise the mind can express itself, but imperfectly. Just carry in your consciousness the thought that the entire physical body is but an instrument through which "MIND MANIFESTS." That every atom of matter is but a vehicle for it. Thus we see how disobeying physical laws can cause disease. You have the key now, so go to work and make your own deductions as to how violation of the rest of the physical laws can cause disease.

A little work on your own account will do wonders for you. Try it.

The *"Mind" force* that causes abnormal pathological conditions is the *same force* that can be turned into producing *normal Real* conditions.

Evidence of the *mind force* working normal and *Real* can be seen in wild animals, who are far removed from the influence of man's unreal thoughts. Here we see evidence of the *laws* of *mind* working free, and unperverted, as was intended by their Manifestor. The wild horse needs no veterinary doctor, the buffalo needs a cow doctor?

Turn to the lower and more primitive races of man for more evidence of the laws of Mind working normally and unperverted. There you will observe the creative Will working through the mind of the entire physical structure, producing a body that is strong, whole and healthy. But let a hand of missionaries, accompanied by one of our learned pathologists, arrive among them, said in the sub-conscious minds of these simple folks will be photographed, picture after picture, of abnormal conditions, and soon these strong bodies will be manifesting all the ills of the flesh.

The mind might be likened to the fertile soil of a

tropical country. If not properly understood; neglected and left unprotected to the mercy of erroneous thought pictures, it will grow the rank weeds of Pathology. But if understood and properly cultivated will grow the beautiful flowers of health. And the reason of this may be understood readily when we understand that the same conditions that are conducive of health furnish the best soil for the growth of the rank vegetation of Pathology.

Without mind matter would be an inert, shapeless, purposeless mass. Let the scientist take the elements of the seed from the matter around him and form it into a seed$_1$ surround it with the proper soil and conditions, apply to it all, the forms of energy known to him, and it will not grow. Why? Because it lacks the third manifestation—Mind and *intelligence.*

Oh, my dear reader, know you that by mind we CREATE, by mind we DESTROY.

We would like to tell, you all about the power of thought. We would like to prove to those among you who do not already know it that thoughts are *Real Things.* We would like to tell you about the, different planes of activity above and below consciousness. How our conscious thinking is only about ten per cent, of the entire workings of the mind. How we are consciously and unconsciously receiving the thoughts sent out by others. How they affect us according to our positive or non-positive mental attitude. More interesting of all, we would like to tell you all about the Instinctive mind, that part of our mind which is influenced by thought, and which has charge of the entire working of our bodies, both in health and sickness, and which also is the seat of all our passions, desires, appetites, emotions, etc., etc. We would like to tell you about the Intellect, which is above the

Instinctive mind in mentation and which is the cause of man's erroneous thinking, until the super-conscious or spiritual plane of mind unfolds, and corrects the delusions created by the intellect. We would like to tell you all about this high part of your mind, which has unfolded into consciousness for the most of you who read these lines, and which is the seat of intuition, which intuitive knowledge enables us to recognize and grasp real truths when they are presented to our minds. Intuition far transcends any knowledge given us by the intellect. Understand, we are not condemning intellect, for without it we could not reason, make deductions, or perceive anything. But intellect alone without guidance from the higher planes of mentation leads us into many dark places. We would like to tell you all about then things, but they are fully covered in the works recommended in the opening chapter of this book.

We think we have told you enough about what disease is. We will now go on and tell you how *"mind"* in action causes abnormal pathological conditions.

We want you to pay *strict attention* to the close correspondence of the symptoms of diseases shown, and the original emotions produced at the time the mind under distressing influence made the mental image, causing the abnormal conditions.

MENTAL IMAGE THEORY

It is not the purpose of a work of this description to go into an explanation of the mental imaging faculties of the mind. That is a subject too large to be handled in a treatise of this kind. Elsewhere in this book we have told you where to get a work giving you this explanation.

We told you also that we were not going to attempt to prove any of the statements made herein. We want you to apply the methods that will be described later on, and they will then prove themselves. Truth is not truth to you until it is realized and experienced by yourself.

Now to go back to our subject. Can mind, forming mental images, cause disease? We answer, *Yes,* and will proceed to show it to you, not by theory, but by citing cases that have come under our personal observation. We will take cases of the class that come to the Osteopathic physician for treatment.

Case 1. Mr.Harry E—, aged 40, came to us suffering from inflammatory rheumatism; almost all the joints in the body were affected, but particularly the joints of the arms and shoulders. Had suffered almost continuously for a period of five years, much of this time being confined to bed.

In searching for a mental image as the cause of the rheumatism, we asked him to tell us of all the accidents that had happened to him from childhood up to the time of applying to us for treatment. He could tell us of none of sufficient importance to have formed an image. So we had to go on and treat him in the regular osteopathic way for about three weeks, when he told us about a fright he had when he was about 35 years of age. This fright proved the cause of his rheumatism, for immediately from the time we received this information and began to apply the curative or antidotal thought he began to make a most marvelous change for the better.

He said that one day he was watching a painter, standing on a jack outside a third-story window, painting, when suddenly the jack gave away, the painter falling face downward to the ground, with his *arms doubled under him,* and was killed.

Is it any wonder that under the stress of this harrowing sight that the will of our patient was as passive, and the mind as subjective, as if he were under a hypnotic spell? The first thing that the instinctive part of the mind did while watching the falling of the painter was to tightly and rigidly contract every muscle in our patient's body, unconsciously preparing to resist the shock of that falling body with the earth. That fall was just as vivid and real to our patient's instinctive mind as it would have been had it occupied the body of the unfortunate painter.

From that time on the mind faithfully carried out the image of contraction in all the muscles, even to *doubling the arms in front of him,* and the only way he could sleep at night was to lie *face downward* with his *arms doubled under him.*

The Osteopathic Physician could have softened up these muscles for the remainder of this man's earthly career, and the mind would have just as faithfully contracted then, and the rheumatism would have constantly reappeared.

Case 2. Mr. John R. S.—, aged 50, came for treatment, apparently suffering from asthma. A careful examination disclosed no conditions that would warrant a diagnosis of asthma. His breathing was most peculiar and distressing. He would puff and blow in paroxysms, like a man who had just finished a ten-mile run. His history showed that about 30 years previously he had undertaken a swim off from Atlantic City to a wreck about a mile and a half from shore. After he had covered about a mile he commenced to get winded, and he realized that he must make the other half mile or go to a watery grave. We will not recount his terrible struggles before he finally reached the wreck, barely able to cling to it, until he was rescued by life-guards.

In this case you see the mind faithfully reproducing unconsciously to our patient, the very urgent desire that the lungs made upon it, for oxygen at the time of that strenuous struggle and still carried out this desire after a lapse of twenty years. This patient is still under treatment, and has been with us for two months. He is very nearly cured. This case requires not only wiping out the wrong mental image, but requires that the mind be taught all over again its function of rhythmic breathing.

Case 3. Mrs. Mary G—, aged 41, came to us in a very much run down condition. One very strange feature about her case which puzzled many medical doctors was the sudden stoppage of her speech. She would be talking freely, as freely and glibly as any woman could, when she would suddenly lose all power of speech. Try as she would, she could not utter a sound for several seconds, when with an effort she would continue the sentence where she had left off. All sorts of brain lesions were suggested by her former physician as the cause. She also had frequent attacks of violent diarrhea. Two years previous to applying for treatment she was a healthy woman. About that time she was sitting on her front porch *talking freely* to her daughter, while her little four-year-old son was riding his velocipede along the outside curb, when around the corner dashed an automobile, stopping barely in time to save the child. The mother was rendered *speechless* for a few seconds, in the *midst of her conversation,* and the fright reacted *on* the bowels, as fright frequently does.

Now, space forbids us giving you more cases illustrative of the imaging power of the mind while it is in a subjective state, from fright or stress of accidents.

We feel that these three cases will be sufficient for any earnest seeker after the truth to go ahead and apply to any case which may come to him for relief and cure. This mental image theory of diseases will clear up and help you out on many puzzling cases. You will *always* be able to trace the *cause,* as well as be able to give the correct instructions to the mind, and guide it back to normal ways.

Remember always that all diseases *begin within the mind*, barring diseases produced by wounds, fractures, etc.

Do not give microbes the power to produce disease. The only power that germs have to produce disease is the power given by a mental image, held in the mind of the germ faddist. This image is easily transferred to the minds of others. The poor little microbe, whose mission in life is to act as a scavenger for us, is blamed for that for which the mind of man alone is responsible.

Woe betide the little child with a sore throat, having one of these germ faddists gaze into her throat. He has such a strong mental image built up in his subconscious mind of diphtheria that he unconsciously and innocently transmits it to the sub-conscious mind of the child. The child may never develop diphtheria, but if it is at all sensitive it may in a year, or two years, or month, or week, develop a true case.

Such is our ignorance that we build up and bestow upon germs a power, which they do not possess, never can possess, outside of the images made in our minds.

Germs are far beneath us in the scale of evolution, and we should always bold ourselves positive to them. The positive and developed man, obeying the laws of nature, has no fear of germs. Knowledge is power. Attain to knowledge and the mysteries of disease clear up as easily as fog before a noonday sun.

I hear you say, "But the bacteriologists have proven beyond all doubt that if they inject diphtheretic genus into a rabbit the rabbit invariably develops the disease."

Let me tell you, my dear reader, the bacteriologist had in *his mind* a vivid image of diphtheria, and it is just as easy to transfer to the instinctive sub-conscious region of the rabbit's mind the image thought of, as it is to that of the child. Dr. Whipple has this to say on the germ theory: "Animals have been impregnated during experiment and the symptoms of certain diseases produced, but later other experimenters have shown that those bacteria were not the ones that caused that kind of disease. What affected those animals?

Note here that the symptoms that develop with the animal are those of "the diseases with which the professor is experimenting." His mind is "full of it." His mind's eye holds the most intense picture of that par-ticular microbe as "the active and real cause of *that disease.*" Rabbits, horses, dogs and guinea pigs all are sub-consciously clairvoyant and subject to the influence of clearly defined thought, by the same laws of *transference of an image* that inhere with human beings.

How much allowance, then, should we make for the perhaps overpowering action of the experimenter's mind on the animal?

None, do you say? Well, let an equally powerful mind give the idea of nothingness to the supposed bac-terial thought and we venture to say that, no disease will develop. It is easy to test. Try it. Fairness and public safety demand a test. We will now go on to the theory and practice of Esoteric Osteopathy.

If Osteopathy is to stand foremost in the ranks of the healing art it must comprehend more than the physical.

Osteopathy must see the power or force that is back of and underlying the Physical body.

Osteopathy *must recognize Mind as that **force**.*

As soon as Osteopathy recognizes *Mind* as the force that moulds and shapes the atoms of matter, either into normal or abnormal states, that moment Osteopathy commands as a MASTER the entire realm of the Physical. Without this comprehension Osteopathy is doomed to wallow in the mire of medical ignorance. It will have its rise and fall like other material schools that are not founded on the rock of truth and *Reality.*

Osteopathy combining its extensive and comprehensive knowledge of physical laws, with the higher laws founded in the ABSOLUTE, will endure without a peer until the end of all Physical life.

If Christian Science, with its glimmering knowledge of the Truth, were to combine that knowledge with such a sweeping knowledge of Physical laws, as does Osteopathy, it would prove itself almost infallible.

Now we will give to you the knowledge which, if practiced will make you absolute master of *dis-ease.*

When the hand that pens these words realizes how *simple* the great truths we are going to give you, will seem and look in cold type, we feel how poor and inadequate is language to express the inexpressible value of the Reality behind the words. But such is Reality. All Real things are simple.

As we have formerly stated, the basic principal of Esoteric Osteopathy is that the disease is a *Mental* trouble—not a trouble in the central mind, but in the

"Mind" in the parts. The theory of the cure is that the thought— force overcomes the "rebellious mind" in the cells and parts, and forces it to resume normal action.

In mind-force healing get all ideas of "matter" out of your mind. You are not using mind against matter, but Mind against Mind. The Will-Mind against the cell mind. Do not *forget this*, for it underlies the whole system of Esoteric Osteopathy. You go after the rebellious "Mind" in the parts—remember that. By producing, or rather re-establishing normal *mental* conditions in the parts, the diseased condition vanishes. We will quote how Ramacharaka, in his *Psychic Healing*, advises to go after the rebellious "mind."

"The Healer directs his thought-force to the 'mind' in the part, and addresses it positively, either by uttering the actual words or by speaking them mentally. He thinks or speaks something like this: Now, *Mind,* you are behaving badly—you are acting like a spoiled child— you know better and I expect you to do better, and act right. You must, bring about normal and healthy conditions. You have charge of these organs, and I expect you to do the work that the Infinite Mind gave you to do properly."

Now there is no virtue in the words used, but this and similar thoughts will give you an idea how to approach the rebellious "Mind." The "Mind" of the part must have pointed out to it just what you expect it to do. You will be surprised how it will obey. Think of a "cross," "pouty" child, one who is "out of sorts." When dealing with the rebellious mind, and like a child you must love it, scold it, coax it, before you can lead it back to right action. Love, of course, must be behind the action, just as in the case of the child. The cell mind is really an undeveloped child-like mind. By keeping this idea in view you will be better able to handle it.

All the while you are manipulating an organ or part, keep telling the "Mind" just what you want it to do. Your manipulations awaken and attract the attention of the "mind in the cells." This explains just how manipulation brings results.

Osteopathic physicians generally have often wondered why continued pressure over a nerve could cause inhibition, whilst a make and break pressure stimulated. Many and various speculations have been advanced as to the cause of this. But not once has it been said that the operator's mind picturing the word "quiet" to the "nerve mind" was the cause of the inhibition, or the picturing the word "action" was the cause of the stimulation. The operator unconsciously thought these word images, and projected them to the "mind" of the nerve centers, under treatment. His very anxiety to bring about the conditions thought of was sufficient to picture to the "nerve mind" what was wanted.

With the light of knowledge, how easily these changes can be produced at will. No longer speculation, no longer doubt, no longer hit and miss, but a *surety,* a *certainty.*

The light of this knowledge explains the results obtained from all Osteopathic technique. In no other way can it reasonably be explained.

Now with a few more instructions we will leave you to work your own way out. We have given you the truths; now go to work and prove them by practice. Don't underrate these truths, because we have told them so simply. We could have woven around these truths much learned (?) theory and conjecture, and thus made a book fifty times larger, but we prefer to present it to you shorn and naked—in a nutshell.

To get quick results, always manipulate directly over an organ or part when giving it instruction, not hard, just sufficient to attract the attention of the "mind." This applies to all but the Liver. Go after it in the usual vigorous manner, bearing in mind that the Liver is a dull, stupid organ, and must be spoken to *sharply* and positively. The Liver cannot be coaxed. It has to be driven like a balky donkey. Of course you understand we are speaking of the "Mind" of the Liver.

The "Kidney Mind" is not quite so stubborn as the Liver, but it needs "talking to" pretty sharply.

The Heart is a very intelligent organ—that is, a higher grade of "mind" than any organ in the body— the brain excepted, of course.

This "Heart Mind" will very readily respond to loving instructions, and is very kind and gentle.

In treating the nerves along the spine, just outline to the "Nerve Mind" in the centers what you expect them to do, and they will carry out your instructions in a way that will please and astonish both you and the patient.

If you have a hard, contracted muscle, just work *easily* over it, telling the "Muscle Mind" to relax, and it will do it in a very short time, thus doing sway with the "back-breaking," strenuous work that you have been accustomed to doing.

Now in closing we will ask you to always bear in mind, that no matter what disease you may be called upon to treat, to remember that it is produced by *imperfect* "mind action." Give your usual Osteopathic treatment, plus the *mental commands* in words suited to the case, and you will be practicing Esoteric Osteopathy in Reality. That is all there is to it.

Above everything else, remember you are talking to the MIND of the organ or part, not to matter. Also always remember that there is no *dead matter* in a live body. Mind is in every part and cell, dead though its matter seems.

Practice will render you perfect, and you will soon

become expert in giving your orders to the mind of the parts. Remember also that "Cell Mind" or "part-mind" does not understand the words you use—that is knowledge they do not possess. But they *do* understand the thought that lies back of the words, and will respond thereto.

Words merely serve to help you to *form your thoughts clearly.* Words are but symbols of thought. Ramacharaka says about this: "A German may give treatment to an Englishman, who does not, understand a single word used. But the cell-mind *does* understand the thought back of the word, no matter what language is spoken. Is not this wonderful? And yet so simple when the key is had. It is the thought, not the word. And yet the spoken word helps the mind to form the thought. *We think in words, remember.* We even dream in words."

Now go to work.

POINTS ON PRACTICE

The Osteopathic Physician should keep his knowledge of Esoteric principles from his patients, not that we advise the use of deception. Not at all. But the general public is not prepared to receive knowledge so far in advance of their usual methods of thought. Dr. Still well says in his "Philosophy of Life" on this very point:

"But we should use caution in asserting that Nature has made its work complete in animal, forms, and has furnished the human body with such wisely prepared principles that the physician can administer remedies to suit the occasion and not go outside the body to find them. Should we find by experiment that the body of man is so wisely arranged by the Deity that the mechanism itself can ferret out disease and purify and keep the temple of life in ease and health, without drugs, we should hesitate to make the fact known. For the opposite opinion has had full sway for centuries and man has by long usage and ignorance adjusted his mind to those customs of the great past; and should he try, without previous training, to reason and bring his mind to the altitude of thought where he can conceive of the greatness and the wisdom of the Infinite, he might become insane or fall back in a stupor and exist only as a living mental blank in the great ocean of life. It would be a calamity to have all of the untrained minds shocked so seriously that they would lose the reasoning power which they now possess. I tell you there is danger, and we must become through a science based purely on the Physical, because then your hopes are pinioned to something that turns to dust. Something that is constantly changing, something that is never permanent."

Do not get "chesty" and "puffed up" and you feeling that you are mightier than your fellows, when success comes to you through this inner knowledge, just always remember that all of us are but poor little "potato bugs" where it comes to *real* knowledge.

Always remember the old saying that "Pride cometh before a fall."

My dear brothers, expressing the hope that you will be successful in your mission of healing—no matter what system you may see fit to use—we will bid you good-bye.

AN AFFIRMATION

"I have within me a great area of Mind that is under my command, and subject to my mastery. This Mind is friendly to me, and is glad to do my bidding, and obey my orders. It will work for me when I ask it, and is constant, untiring, and faithful. Knowing this, I am no longer afraid, ignorant or uninformed.

The "I" is master of it all, and is asserting its authority. "I" am master over Body, Mind, Consciousness and Sub-Consciousness. "I" am "I" — and "I" am Spirit, a fragment from the Divine Flame."

—

Ramacharaka.

Part 3

Epigenetic Noetic Virus

Epigenetic Noetic Virus - is there a cure?

An examination of the epistemological and metaphysical postulates underlying the contrast between traditional and alternative medical science.

By Zachary Comeaux DO

Preface

"Ha!" you say. "What could the title mean? Is this a joke?" And indeed, the title is a bit of a spoof on the perennial and serious quest to get to the root of our existence. *Epigenetic* refers to the prospect of activity more fundamental than genetic initiation or determinism. *Noetic* refers, in the philosophical context, to ideas, or consciousness. I used the term *Virus* to describe this quest to understand as something which, once it infects us, is virtually incurable. I am sympathetic but merely want to recognize the fundamental difficulty in exploring these themes.

The essay takes a cursory long view of the issues and authors involved in the subject of the current book – what is life. In some senses it parallels the other two but carries it forward to the contemporary debate, after the great Human Genome Project. There are a myriad of sources to read. I include just a few critically placed opinions. The pattern follows the theme laid down by the founder of osteopathic medicine, to create a system based on an all-inclusive view of the human person, in Nature.

Introduction: epistemology and metaphysics implicit in medicine.

In the medical encounter, symptoms are analyzed in the context of one's view of the world. There is a challenge to blend the paradigmatic view of the patient, the culture, and medical science. The physician must develop effective consensus among the three based on what we know, and how, and what can be done or predicted. Most physicians are trained in the reining paradigm of materialistic, mechanistic biophysiology with the sole underpinning of biochemistry rooted in Newtonian physics. Hence the dependence on pharmacologic intervention. Evidence based medicine, compatible with insurance industry interest in the idea of management of populations of "insured lives", is the attempt at the application of this scientific method and is based on statistical rather than individual outcomes.

The materialist reductionist position, and its implied metaphysical stand, represents the institutionalization of a response of the science community at one moment in the material-determinist /vitalist argument. Dating from the era of Aristotle and Democritus, the last serious competition of these apparently dichotomous views of reality and life occurred during the latter half of the nineteenth century and beginning of the twentieth. [1] Vitalism was challenged by the successes of material interpretation reinforced by the successes in mechanistic technology and chemistry. The Flexner Report (1917) mandating substantive reform and standardization of medical school curricula galvanized the then current view and role of physiology. [2] Recent discussions by biologists repeatedly refer to the input of vitalism as of historical interest only.

However, the streams of thought and discussion from which these paradigms derived have all evolved significantly. Medical science has progressed, in the era of DNA technology, into advanced levels of detail in observation and manipulation of the organization of life processes including embryogenesis, implantation and cloning. Concomitantly popular culture has suggested intercultural and existential approaches to justify "alternative" and "complementary" medical practices. These are often based on different postulates more akin to materialistic or animistic vitalism's view of the nature of life and of the patient. They suggest that there is more to the living organism than the progressive expressive of genetically programmed events and traits. Biophilosophy, additionally, has continued to evaluate the ontologic and epistemological postulates of scientific and popular views of reality, nature, and the person and to attempt to refine a systematized scheme for describing life.

One of the problems of science is the accounting for species similarity yet individual variation in the formation of each living organism. Philosophically, epigenesis presents the question "Is material reality sufficient to describe the entering into existence, or persistence and nature, of a living individual?" Historically it has driven the argument for the implication of non-material life forces in explication certain aspects of human existence and biologic process generally. Clearly this impacts on one's postulates for a scientific method of health science and the selection of therapeutics. In the materialist paradigm, a force does not exist if it can not be observed, quantified, and predict. Yet there is always a limit to what one can explain or predict.

How do we include the indescribable in our system? Does it warrant inclusion?

The current interest in psychosomatic effects and consciousness studies is one attempt to include and legitimize variant approaches. Biophilosophers, scientists with a philosophical bent [3], and physicians [4, 5] also challenge this criterion, materiality as scientifically understood, as definitive of living processes. Some of the support for postulating effects of non-Newtonian forces was first hypothesized by von Baer, and later measured by Nuccitelli and others, in the form of bioactive electromagnetic fields. If relevant, can health be restored to the ill more effectively by taking these forces or influences into account? How are they embedded in the fabric of life? Do they warrant inclusion in a scientific paradigm?

Embryologists on the other hand have begun to appropriate the term *epigenetic* with new nuances often to reinforce a materialist cosmology, or determinist world view. [8, 9] Although genetic control of protein synthesis influences much of early development, some embryologists feel embryogenesis is dependent on undiscovered cytoplasmic forces. Does the concept *epigenetic* have any useful meaning at all? Is it simply an anachronistic resurrection of vitalism, a dragon long slain, or is it a helpful tool in the refinement of bioscience? Or are non-physical (by Newtonian definition) factors involved in the fostering of life and health? Should the basis for alternative therapies be reconsidered?

This paper is an attempt at summarizing the principal positions of the classical period, the turn of the last century, and the beginning of a new millennium on the question "what is life" from the materialist and vitalist points of view. What I want to elucidate is the cyclic process by which one's premises or postulates form the criteria by which the validity of the outcome is assessed. This issue is important whether one is a patient expressing a complaint, a physician making a diagnosis, or a medical researcher formulating a hypothesis, or a committee developing health policy. It shades the character of what is true observation of nature and an objective standard? It interjects a dimension of subjectivity into every "objective" process and activity.

The Ancients: dominance of deduction

The term *epigenetic* may not have been used by Aristotle, but as Dreisch points out, the sense in which it is used today is contained in Aristotle's analysis of conception, the formation of life. Aristotle attempted to define effects on a conceptus, or individuation of a new living creature, that were not due to the substrate of the ovum and sperm. In applying rational analysis, he suggests a transition from potentiality to actuality (entelechia) to describe the character of formation of new life. He develops the role of a soul to accomplish this transition, yet the soul is no more separate from the body than in the case of wax and its form. [10, p.18] In contending with opposing views, including the Democitean position that matter begets matter self sufficiently, Aristotle postulates the soul as an intrinsic, necessary causal entity. Causation involved the material cause of the substrate, and efficient cause or force. Moreover individuation relied on a source of the formal cause, (the morphology). Aristotle included in his universe final cause (teleology). These required the guidance or input of something more, something extra-material. [11, p. 36]

So begins an argument over whether, in trying to understand the character and activity of life, conceptualizations represent real divisions in nature, and separate moieties and agencies, or were artifacts of rational activity. If separate, were they intrinsically or extrinsically driven or defined? For centuries, rational constructionism, and deductive reasoning became the basis for systematic philosophy including biology, the philosophy of life science. Reinforced by the cosmology which supported man's special place in the universe, it became institutionalized as the basis of social and political order, veiling the original philosophical bias which selected it as the dominant paradigm of science and knowledge. In essence, power became a valued commodity over intellectual honesty.

Cartesian Revolution: an inductive reaction

In reaction to the absolutizing of reason, Descartes and others defending the primacy of the observable phenomena of the world, and life, in mechanistic terms. We have no time here to review the whole progression of thought. Descartes is most noted for the conceptualization of the mechanical character of physical reality and the required separate realm of animation needed for living creatures. Recognizing the inadequacy of this division, in attempting to make independent the sphere of natural observation and science, Descartes' dualism still recognized the activity of a Creator and the free will of man. [12, p. 375]

Unable to develop the primacy of one of these spheres of reality, a person must operationally bracket each domain, operating separately in each.

Nineteenth Century: Materialism vs. Vitalism

The mid-nineteenth century, into this century, reflects a renaissance in the philosophy of life science. For simplicity we will concentrate on the highlights.

The most commonly recognized influence on nineteenth century biophilosophical thought was the work of Charles Darwin. However, Darwin's thought emerged from a rich context of scientific debate. Pre-Darwinian attempts to reconcile the organized character of embryology of the individual and group conformity as species reached its height of sophistication in Germany in the eighteenth and nineteenth centuries. Apparent purposefulness in the variations in development and the reaction of the organism in an organized fashion were noted as a theme, recognized as teleology. Beginning with the Kantian premise that organization in the individual presupposes an original state of organization, the school of *romantische Naturphilosophie* proposed a variety of frank teleologic approaches.

The embryologist Karl Ernst von Baer [13], attempted to avoid extrinsic vitalistic implications postulating the concept of Type to describe the orderly progression of spatial differentiation of tissue in common patterns of development of the species. It might be noted that much of the categorization of events, tissue types, and functional embryonic morphology currently used derive from this period. Von

Baer describes Type as the "relational position of the organ elements and organs of the embryo... the expression of certain primary conditions in the direction of the particular relations of life." He presages the concept of induction of pluripotent tissue. Elaborating on the Kantian concept of predetermined order, he introduces the concept of Keim or Anlagen to describe the ancestrally conveyed potentiality of the individual. (Here he presages genetic transmission.) Von Baer's analysis comes close to Aristotle's final causation, discarded by contemporary biologists because it conflicts with the premise of time as sequential and the problem of reverse causation. He raises the question of a force that could be responsible for the structuring of the organs in the tissue and consistent with the universal force of nature. He hypothesizes an electromagnetic agency.

Throughout his synthesis, von Baer insists that whatever forces precipitate and regulate epigenesis cannot exist independently from organized matter. [8, p. 86]

Herbert Spencer claimed to predate Darwin in the theory of evolution, although his mechanism for evolution relied more on physics than on intracellular chemistry. Recognizing philosophy as "Completely unified knowledge", Spencer developed a classic systemic philosophy, psychology, sociology based on the relationships of matter, motion and force. [14, p. 113] Building on a base of Newtonian physics, Spencer postulate alternative repulsion and attraction as developing the stabilizing forces for apparently continuous states and relationships. Biologic processes, including consciousness, are refined levels of such interaction. Change, including biological evolution, is the succession of changes accomplished by interactive forces and resultant inertial states. Notably lacking in his systematization are specific examples of physiologic function. In principle he stated that even conscious process follows the laws of mass attraction/repulsion, inertia, and oscillatory motion.

Charles Darwin is generally attributed to applying Mendelian genetics to the natural world in a systematic way. One should note that these hypotheses by a naturalist occurred in the context of a continuing scientific debate and that others had postulated the evolutionary process, suggesting mechanisms. Darwin proposed that natural selection followed successful random meiotic and mitotic events and that mutated characteristics were retained. Lamarck before him had made similar suggestions regarding the individual. Darwin's contribution seemed to be postulating this process of change to definition of species. Environmental challenge to mutations elected the character of the species by selective survival success. His definition of the problem of individual epigenisis was one of genetic expression with mutation as the sole source of variation. With refinement, it serves as the basis of modern genetic theory, with the addition in this century of the ability to observe and manipulate cellular transactions involving bio active protein. However, it includes no formal ontologic or epistemologic reflection, accepting a materialist world view.

Hans Dreish, an embryologist, reviewed the observations and theories of both the physicochemical approach to science and the vitalist school. Dreisch wrote during a period in which microscopic evidence was beginning to discredit the view of spontaneous generation of worms, previously held as a proof of epigenetic activity. However, even in the presence of microscopic material effective activity, Dreish defended a need for formative and final causation as introduced by Aristotle. Most notably this was required to describe the idea of form, or overall organization of the organism. He used Aristotle's word *entelechy* as the agent of actualization. However the extra entelechy responsible for induction of the morphology from the molecular substrate was, as Aristotle's soul, not a separable entity. He dismisses a Cartesian dual world view of some prior "naive vitalists" and postulates that there is another level of logical relationship necessary to characterize living organisms over simple physical objects. This is necessary to confirm our conscious experience of enduring and becoming [10, p.188 ff.] of autonomous life processes. Physical reality, at any one time, entails a homogeneous distribution of possibilities that may be transformed into a heterogenous distribution of realities" [10, pp .203-4) Entelechy entails the suspension of possibilities for the realization of one specific form but, "it is on given, preformed, material conditions that entelechy depends." In itself it is "neither an energy, nor a material substrate, but a non-material, non spatial agent".

Henri Bergson, an early a student of Spencer, built on the postulating of a cosmology, using the experience of perception as a beginning point. Coming from the initially critical view of the limitations of the representation of reality as the object of rational analysis, Bergson parallels Dreisch in characterizing life as experience of becoming, possessing the quality of duration and "growing old." Rational categorization cannot capture what is real since "reason is created by life; how can it reflectively comprehend life". "Life evolves before our eyes as a continuous creation of unforeseen form"; the intellect "instinctively selects in a given situation whatever is like something it knows. .... Science carries this faculty to the highest degree of exactitude and precision but does not alter its essential character." [12, p. 35] He further challenges the concept of time as an independent variable, intimating that this conceptualization does not correspond to the true duration in nature.

Bergson, sometimes characterized as a proponent of a dualistic, animistic vitalism proposing an extrinsic "elan vitale" is rather attempting to maintain the unified experience of reality in consciousness as the gold standard for contact with nature. This *vital impetus* is simply acquiescence to the effort to define the movement of life in mechanistic terms; it is postulated as the source of evolutionary inertia. {12, p. 97] However once suggested, Bergson makes clear, "Mechanism would consist in seeing only the positions. Finalism would take their order into account. But both mechanism and finalism would leave on one side the movement, which is reality itself." [10, p. 101]

This milieu of philosophical reflection on the origin and operative principles of life contributed to the intellectual climate of American medicine at the turn of the century. For some, the assent to a chemical/physical model of human physiology developed into a materialist deterministic approach which emerged as allopathic, or contemporary mainstream medicine. This faction, through the agency of the Flexner Report, effected the character of

American medical education and its ancillary sciences. For others, the postulating of bioactive electromagnetic and other forces reinforced the belief in Spiritism. Similar experiences were responsible for the evolution of homeopathy and eclectic medicine. On the frontier, A.T. Still M.D. reflected on the extra material influences on structure and function of the individuals who he saw as tri-partite by nature, possessing a physical body, mental body and body of mind. [15, p.16])

Still's legacy is the Osteopathic profession, founded mostly on his ability to cure disease by restoring normal relationships to functional anatomic parts through manual manipulation. He began with a Cartesian style Creationism in which the body was seen as a machine with God as the mechanic. However, in his mature thought he postulated extra material yet intrinsic causation of life process involving an agency he calls *biogen*; "We see the form of each world, and call the united action biogenic life. All material bodies have life terrestrial and all space has life, ethereal and spiritual life. The two, when united, form man." [15, p.251]

Twentieth Century; Metaphysics versus science and the need to reflect on premises

The twentieth century brings a progressive ability for more detailed observation and manipulation but the persistent problems remain in defining an organizing principle to life, and dealing with issues of progressive individuation of species forms.

"Yet the fact that some ontologies are wrong or useless does not render all metaphysics objectionable, every human belief and action involves some metaphysical presuppositions. Thus as has been remarked many times, and rightly so, than antimetaphysician is just one who holds primitive and unexplained metaphysical beliefs."[11, p. 3]

This however, is not the spirit in which much science was practiced at the beginning of the twentieth century. A culture intoxicated with a series of mechanical advances, an expansion of chemistry, and physics saw unlimited application to the life sciences and medicine. At the forefront of theoretical physics, Erwin Schrodinger remarked:

"How can the events of space and time which take place within the spatial boundaries of a living organism be accounted for by physics and chemistry?...The obvious inability of present day physics and chemistry to account for such events is no reason at all for doubting that they can be accounted for by those sciences." [16]

Schrodinger, a revered physicist, saw the potential transfer of concepts from physics into the field of genetics and was enthralled at the potential for defining the physical substrate of life. He inferred a parallel hierarchy of ontogenic significance on the hierarchy of structure of gross, microscopic, molecular, anatomic and subatomic observable phenomena. He conjectured that bioactive molecules may represent the "aperiodic crystal which in my opinion is the material of life". He reflected on the energy thresholds between biomolecular energy stated and analogized this to the mechanics of genetic mutation. He conjectured that, "it emerges that living matter, while not eluding the 'laws of physics' as established up to date, is likely to involve other laws of physics hitherto unknown, which, however, once they have been revealed, will form just as integral a part of this science as the former."

Indeed he went on to postulate and defend the concept of "negative entropy" as a characteristic of living organism, the ability to extract order out of the environment. He remarked that this process consists in events whose regular and lawful unfolding is guided by a "mechanism" entirely different from the "probability mechanism" of physics, a fact of observation". In so doing he attempts to postulate an escape from the Aristotelian and Kantian logical problem of the origin of order from order based on observation, akin to the concept of emergence as a property of living systems not simply a function of progressive knowledge of an expansive yet constant universe, as discussed below.

A counter view, attempting to systematically bridge the epistemology/ontologic gap is the work of Walter Russell. Russell, [17] a philosopher poet inspired by Herbert Spencer attempted to modernize the latter's work by emphasizing the oscillatory property of matter. He postulated rhythmic motion as the basic entity of life and argued that matter, as it expresses itself in time is a derivative phenomenon. Russell developed this theme reinforced by the rhythmic regularity of the chemical periodic table, suggesting yet to be discovered elements. He postulated that concentric levels of harmonic effect, in neo-Pythagorean fashion, were responsible for mental, emotional and spiritual reality and were potentially able to be influenced, as was physical matter, if one knew and practiced the principles of Rhythmic Balanced Interchange. His hypothesis was that an entity analogous to light (he uses the word Light) is the highest known element and has motion characteristics. Physical reality is a lower level emanation of reality at this higher level. (His language parallel's Einstein's fascination with the special significance of secret of life.) Writing in a poetic aphoristic style, one could dismiss his thoughts as naive, or begin to make connections with then contemporary physics. He was influential with some leaders of manual medicine, including R. Fulford and W.G. Sutherland, who observed endogenous physiologic rhythms as coextensive with Life.

Two incisive analytic views clarify the apparent disparity between the materialist and vitalist labels. J. H. Woodger [18, p. 229 ff.] went beyond the labels to the substance of the positions. He classified views as either *dogmatically* vitalist (or materialist) which involved metaphysical implications or *methodologically* vitalist (or materialist) which avoided ontologic considerations which yet had to be considered. He then went on to assess individual positions according to their intent to answer either ontologic or practical questions. A methodologic materialist would not then assert that the human person was a machine. He would only state that there were many situations in which for the sake of problem solving, the human person could be considered a machine. A dogmatic (metaphysical) materialist would hold that the human person was in every sense nothing more than a machine, operating under the known laws of chemistry and physics.

Epigenetic Noetic Virus

This view is complemented by Karl Popper [19] in his erudite analysis of scientific methodology. Popper observes that the reductionism exercised on the data of biological observation are sometimes successful, sometimes unsuccessful yet always leave a "residue", something left unexplained. As an example he cites the monumental utility of Newtonian explanation of physical laws but recalls the need for later developments to refine its application on the level of subatomic particles. This, he reflects, was not necessarily because of sloppy thinking, but is part of the inherent problem of rational reduction as can be demonstrated in the world of mathematics by the problem of irrational numbers. He traced this process through the evolving field of physics and the infinite sequence of "final discoveries" and subsequent exceptions. Popper's own attempt at cosmologic reductionism suggested reality can be divided into three tiers. World one is the world of "physical matter, fields of force; world 2 is the world of conscious and subconscious experience. World 3 is the world of the spoken word." Each world partly autonomous yet related to the other three. Popper's analogy has been challenged by materialist reductionists who accept the reality only of the first. This can be viewed as a proof of Popper's theory of the limitation of reductionism or a discrediting of his thought in itself. However, Popper brings home the issue that the veracity of science is rooted in the epistemological/ontologic problem, which he refers to as the *residue.*

The ultimate application of reductionism to biologic process in the last hundred years has been the progressive discovery of molecular genetics and proteomics. Michael Ruse [20] explored this field for its potential as the occasion for reducing all of biology to physics and chemistry. He is clear that genetic interaction can be expressed in the form of mathematical symbolic logic, however in order to deduce a biologic system to axiomatic form we need to decide what are the objects of study. Are they theoretical or observable entities? [20 p.11] Clearly our assessing cell process, in which genetic behavior occurs, involves an abstraction as object, certain aspects of a continuum of activity. Likewise, when dealing with the ideas of goal directed systems, he attempts to distinguish concepts of causality, versus adaptation versus systemic function. The discussion is erudite yet he admits that many biologists while dismissing vitalism, "feel that some facets of the biological world will forever elude a purely physico-chemical analysis." [20, p. 209] In his later revisitation of the subject [21, p. 49], he accepts biological teleology, or purposefulness, as a reality which, beyond prevention of absorption of biology into physics, may actually change physical conceptions themselves.

In another attempt at a systematic reconciliation of physics and biology Mahner and Bunge present a different view of the philosophical underpinnings of science. [11]The authors begin by making several postulates that metaphysically define the world in materialist terms. The world is made of things which are material objects. They call this systematization a methodologic dualism (material things and immaterial constructs). However, they summarily dismiss the existence of constructs. In so doing they both dismiss Popper's "world 3" becoming in Popper's words disguised dogmatic or metaphysical materialists. Whatever does not fit in the symbolic logic description of the universe is ignored. In methodical fashion they review the history of biologic reflections, criticizing each stand in light of their postulates. The key however, is the fact that the process necessarily begins with postulates, representing subjective bias.

Curiously, in reviewing the concept of emergence or emergent properties of matter, they contrast the thoughts of the Schrodinger style reduction with various other expressions. The concept of emergent processes is one means of explaining the observation of apparent "qualitative novelty".

A reductionist view would consider emergence an illusion caused by refinement of previous capacity to observe and measure, without ontologic basis. This seems to conflict observation of biologic phenomena. They postulate a metaphysical basis for emergence. They will not postulate a mechanism beyond proximate sequential concrete effects, or functional relations among properties. They categorically dismiss the concept of "ultimate cause", Aristotle's formative or final causalities. By textual juxtaposition they intimate an involvement with self-organization in biology but do not elaborate.

Emergent Paradigms- Theoretical reorientation

In a detailed attempt at blending the intention of von Baer and his colleagues in investigating the ontologic and epistemological principles with contemporary embryology, Soren Lovtrup begins by reviewing the basis for scientific observation and theorization. He cites Woodger in the need for science to "discriminate events in the 'real' world and to invent abstractions for their classification". [8, p. 5] He clarifies that epigenesis has been used both to describe early ontogensis on the material level, and to describe the formative process of the individual in development from embryo to adult. In any case, he cites Woodger's characterization of epigenesis as "a spatio-temporal happening", reflecting cellular processes. The body of the work is a detailed presentation of cell theory as a basis for analysis of development. He frequently acknowledges the consensus view, that development depends on protein synthesis reliant on DNA transcription. However, numerous other effects are due to characteristics of cell line, physical characteristics, polarity and environmental factors. What governs the overall organization? A variety of genetically encoded potential organization patterns are eliminated by the expression of one potential pattern. He postulates that individual epigenesis is the result of individual intercellular physical forces, resulting in sequential induction of further intercellular events. This is the basis of the final form, the morphogenesis of the individual. In other words, much is still not known, nor can it be generalized, hence the problems with cloning.

In reviewing the process in lower chordates, P. Nieukoop [9] suggests that induction of pluripotent tissue requires a signal. He concurs with Lovtrup that the cell is the smallest unit of autonomous function. However he cannot agree that the autonomy is absolute, and certainly is not dependent solely on genetic inscribed signals.

"Unfortunately, we know little about the regulatory mechanism involved in nucleocyoplasmic interaction. There seems not to be support specifically in basic proteins such as histones, nor in polymerases to account for complex differential activation and repression of genes during development and differentiation."

Rupert Sheldrake [3] suggests another conceptualization of the problem of epigenesis. Confronted with the problem of apparent emergent cell behavior, goal directed behavior in development, regulation and regeneration, he challenges the physical materialist view.

"Indeed the properties attributed to genetic programmes are remarkably similar to those with which the vitalists endowed their hypothetical vital factors; ironically, the genetic programme seems to be very much like a vital factor in a mechanistic guise." [3, p. 22]

Focusing on the question of pattern, or form, Sheldrake postulates the relationship of formal cause through the vehicle of the *morphogenic field*. [3, p. 70] This is distinct from a subtle energetic field potentially measurable by physicists. The manner of effect is one of resonance by which the effect is somehow transferred. He describes these activities as an ontogenic extension of the nature of bodies. His 'proof" is the consistency of effect of organization noted between physical entities at all levels of complexity, an emperico-deductive proof. By inference this represents an attempt to reformat our perception of reality and also to attribute to matter another level of self organization.

Soul of Osteopathy

The concept of self organization of matter, implicit in the subtle intercellular embryogenesis of Lovtrup, and the theoretical cosmology of Spencer and derivative theories challenges physical measurement. Reflecting on nature as process or activity more than form, J. Kelso [22] related patterned activity to a dynamic systems model. Citing A.M. Turing he quotes "Most of an organism most of the time is developing from one pattern to another not from homogeneity into a pattern." Using concepts such as stable and unstable systems and conditions precipitating bifurcations in activity paths, in dynamic patterns, made mathematically representable using attractor patterns from chaos math theory, Kelso applies these ideas to living, dynamical, "open" systems. In so doing, he implies that the materialists are correct in attributing form and change to discrete sequential interactions between discrete material entities, however he attributes inherent dynamical effects not covered by simple Newtonian interactions. He postulates to matter a self organizing capacity not usually credited by materialist approaches.

In human biology the centrality of consciousness is a problem to confront in definition of life process. Kelso and Sheldrake hypothesize that consciousness, an effective force, is a subset of their dynamical systems. There is some room for speculative application of this theory in light of Hebb's concepts of resonant cell assemblies as a unit of neural processing. Can will, or intention, then influence physical outcomes by effecting neural activity? In a novel series of experiments, R. Jahn [23, 24] at the Princeton Engineering Anomalies Lab showed slight but repeatable bias in the results of randomly generated binary sequences toward the   pre-stated operator intention. Similar results were obtained regardless of physical proximity. Blinded retrospective intention showed similar results. If such data on living systems are valid, the work of Bergson in challenging the convention of the absolutely sequential nature of time may be justified. If inert objects can be effected by intention, what of living biomolecular process? The challenge of measurement and demonstration of these theories remains.

Lessons: Scientific and Clinical

Although the above is but a sampling of the earnest reflection on the methodology of science, and its implied or postulated underpinnings, it should by now be appreciated that the nature of observation and experience requires an ontologic and epistemologic basis for analysis. As noted by Popper, any material analysis leaves a residue, some aspects of reality that do not fit the rational reduction. This is the main point in Bergson's critique. As noted by Kelso and others the problem is implicit in the quest to explain life since Life is an "open system". Any reductionist meteorologically must close the perimeter of observation.

Closing the perimeter is the developing of postulates. The basis for the choice seems, on review, to be a matter of one's intentions in face of the problem to be solved, tempered by experience. Axiologic or value driven considerations may influence development and acceptance of postulates. Experience however is an individual affair. Consequently, the derivation of a rational system of science or philosophy would appear to be an individual affair. To save us from utter subjective phenomenology, one can repostulate the objectivity of the "world". However, to arrive at this point one must recapitulate the epistemologic arguments of the last three millennia. In this context we reiterate "what is life?" Is it materially based or are there extra material influences which shape physical reality? Is material reality self organizing? Is it genetically preprogrammed? What is the source of the program? Schrodinger suggested it is just a matter of time before the gaps in knowledge are filled by observation. Spence, the absolute systematizer, claims,

"The conviction is reached that human intelligence is incapable of absolute knowledge. The reality existing behind all appearances is, and must ever be, unknown."

Both Bergson and Dreish lose the object observed in an infinite dissolve, looking at the reflection of reflection, as an inherent condition of human perception. Life is real but known only as it is held in the transient stream of consciousness, the epistemological problem. Hence their emphasis on the aspect of movement (Bergson) or becoming (Dreisch). The reality of experience however moved them then to postulate an active entity responsible for the life process. Very close to this is the latter work of Sheldrake and Kelso, again based on observation, the emphasis on process. Is this inconclusiveness simply a result of "fuzzy" non-rigorous thinking?

Mahner, the realist materialist stated in the chapter *What is Life*:

"Yet, even when approached scientifically, a satisfactory answer is not to be expected as long as living beings were studied either on their own level (holistic approach) or as physical systems devoid of emergent properties (reductionism), and in either case apart from their history."

Yet history is never done and is apprehended only as progressive observation. Hence he admits that any postulating and systematization represents premature closure of an otherwise open (temporally) system. He capitulates, after an exhaustive philosophic methodologic review, that his synthesis is "a systematic, if only partial, philosophy of biology."

Lovtrup though optimistic recognizes the meager results of the last hundred year process in attempting to ascribe "modi operandi" to easily observable embryogenetic processes, yet does not admit to the epistemologic disclaimer. He refers to the challenge to "some kind of resolution (reduction?) - experimental as well as intellectual- is necessary before it is possible to disentangle the workings of the forces involved." Clearly he admits we are not able to define life process at this point, yet he shares Schrodinger's faith in the scientific academic pursuit of truth.

It would appear then that, despite heroic and assiduous intellectual effort, the problem of what is life, and how to make comprehensively successful interventions is based on value judgements; one is involved in a closed loop. One observes nature to assess its conformity and variation from "known" laws. Yet derivative of laws depends on experientially limiting the field of study. A certain power of predictability comes from limiting extrapolation from experiences which seem reproducible by a consensus of the observers in one's discipline. Further predictive power come with consilience between fields of study. This reductionism does not encompass the whole of reality and one often is taught over time that a significant portion of the system has been left out, Popper's inevitable residue.

In any case, Whitehead's caution of years ago seems pertinent:

"If science is not to degenerate into a medley of ad hoc hypotheses, it must become philosophical and enter into a thorough criticism of its own foundations." [25]

Practical Application: "What am I to do, Doctor?"

In our paradigm, physicians are looked at as sources of the truth about human biology and practical consequences. One needs to know the derived "standard of care" and the physiology and epidemiology on which it is based. One can realize that a certain amount of error and obsolescence is built in because of the methodologic bias of empiric reductionism. But beyond that, other factor's effect patient outcomes.

Significant study and long consideration of the reality of psychosomatic aspects of illness and health, including the placebo effect, suggests to me that emergent properties of the central nervous system and the translation of the effects of intention, or thought in general, into cell mediated physiology is not so far fetched. Clinically this gives me much respect for the impact of patient expectation, attitude and patterns of activity and thought. Risk factor analysis overlaps with this area of impact to a significant extent. Clearly good rapport between physician and patient, trust in a system, or trust in an "alternative" therapy may relate to human physiology by more routes that pharmico- or physio-chemical influence.

Several of our scientists and theoreticians in this review have cited the primacy of the "temporo-spacial" uniqueness of this individual. Each patient is an open ended system. In some ways they will conform to, in some ways diverge from, a normal statistical population. Clinical judgment is best tempered by consideration of the patient as a potential outlier. In any case a position of consideration of the patient and physician as co-investigators, with variable degrees of commitment to the project of health seems realistic. The physician might be cautioned to consider the idiosyncracies of this patient's particular body system and processes while pursuing algorithmic thought.

A panel of experts were once pooled according to how they made clinical judgements. They indicated that their diagnosis was often fairly final after two minutes of patient-physician exchange. When queried as to the algorithm or other decision pattern used, they cite "intuition", certainly based on the summation of experience. What is routinely scribed to be subjective and therefore less valid, may be the mainstay of expert interaction.

I would sum by saying simply that in face of scientific pressure to predict (including the Human Genome Project) and managed care application of actuarial figures to patient encounters, that history is an open ended process. Each patient is an unfinished history and therefore an open-ended system. And reticence at dogmatic prediction of outcome or absolute definition of clinical condition is prudent. This extends to rigid conformity to a dogmatic set of practice protocols. Clinical guidelines are a help in clear thinking but an epistemological disclaimer regarding the certitude of knowledge ought to be translated into the vernacular of the patient. Treating, health, illness, and patient as an expression of biologic process and relationship is more true to life and Nature than diagnostic labels and absolutely correct procedures.

In other words, the issues and questions implicated in the use of the word epigenesis suggesting more than materiality, first used by Aristotle, remain current and unresolved. Like a virus they are still with us, resistive of cure, so far. They challenge each of us in defining human health to consider Life as an open system. In this context one must do research, live, and practice medicine with an open mind.

The founder of Osteopathic Medicine, Andrew Still MD would direct us in diagnosis and treatment to "find the unnatural and return it to the natural." What seems to be lacking in precision may be wisely compensated by the comprehensiveness of the approach.

References:

1. Montalenti, G., *From Aristotle to Democritus via Darwi"*, in Studies in the Philosophy of Biology, Ed. F. Ayala, T. Dobzhansky , U. of California Press, 1974 pp. 21-4

2. Suter, E., M.D., Mandin, H., M.D. *The Science, In the Education of Physicians*, in Basic Science Educator, Vol.8, No.1 and 2, pp. 7-9

3. Sheldrake, R., *A New Science of Life; the hypothesis of morphic resonance*, Park Street Press, Rochester, Vt. 1981

4. Dossey, L., Healing Words, the power of prayer and practice of medicine, Harper, San Francisco, 1993

5. Fulford, R., Dr. Fulford's Touch of Life, Pocket Books, N.Y., 1996

6. Von Baer, Entwickelungsgeschichte, Scolium V, p. xxii Konigsberg 1828

7.Nuccitelli, R., *Endogenous Electric Fields in Developing Embryos"* in Electromagnetic Fields, M. Blank ed., 1995, pp. 55 ff. p. 109,

8. Lovtrup, S., Epigenetics; A Treatise on Theoretical Biology, Wiley, London, 1974

9. Nieuwkoop, P., The Epigenetic Nature of Early Chordate Development, Cambridge U. Press, 1985

10. Dreisch, H., The History and Theory of Vitalism, Macmillan and Co. London, 1914

11. Mahner,M., Bunge, M., Foundations of Biophilosophy, Springer, Berlin, 1997,

12. Bergson, H., Creative Evolution, Henry Holt, N.Y., 1911

13. Lenoir, T. The Strategy of Life, teleology and mechanic in Nineteenth Century Germany, D. Reidel, Dorchester, 1982

14. Spencer, H., First Principles, Burt Publishers, NY. 1880

15. Still, A.T., M.D., The Philosophy and Mechanical Principles of Osteopathy, Hudson-Kimberly, Kansas City, 1902

16. Schrodinger, E., What is Life, the physical aspect of the living cell, Cambridge Press, 1967

17. Russell, W., The Universal One, University of Science and Philosophy, Swannanoa, 1926

18. Woodger, J. H., Biological Principles, a Critical Study, Routledge and Kegan, London, 1929

19. Popper, K. R., "Scientific Reduction and the Essential Incompleteness of All Sciences", in Studies in the Philosophy of Biology, Ayala, F., Dobzhansky,T. eds., U. of California Press, Berkeley, 1974,

20. Ruse, M., The Philosophy of Biology, Hickman and Co., London, 1973

21. Ruse, M. The Philosophy of Biology Today, State University of N.Y., Albany, 1988

22. Kelso, J., Dynamic Patterns, the self-organization of brain and behavior, MIT Press, Cambridge 1995

23. Jahn, R., Margins of reality, the role of the consciousness in the physical world, Harcourt Brace, 1988

24. Jahn, R., "Correlations of Random Binary Sequences with Pre-Stated Operator Intention", J. Scientific Exploration, v. 11, no.3, pp. 345-367, 1997

25. Whitehead, A. N., Science and the Modern World, Cambridge 1927 p.20

Lightning Source UK Ltd.
Milton Keynes UK
UKOW03f0945180514

231866UK00001B/14/P